T0370032

# AMAZING WORLDS

OF SCIENCE FICTION AND SCIENCE FACT

# AMAZING WORLDS

## OF SCIENCE FICTION AND SCIENCE FACT

KEITH COOPER

REAKTION BOOKS

Published by
REAKTION BOOKS LTD
Unit 32, Waterside
44–48 Wharf Road
London N1 7UX, UK
www.reaktionbooks.co.uk

First published 2025


Printed and bound in Great Britain by Bell & Bain, Glasgow

A catalogue record for this book is available from the British Library

ISBN 978 1 78914 994 4

# CONTENTS

# 1

# STRANGE NEW WORLDS

James Tiberius Kirk said it best, declaring at the beginning of each episode of *Star Trek* that the mission of the good ship USS *Enterprise* was to 'explore strange new worlds'. Astronomers have been exploring strange new worlds for real since 1992, when the first bona fide planet to be found orbiting another star beyond our own solar system – an extra-solar planet, or 'exoplanet' for short – was discovered. Even now that first discovery is relatively recent. Yet science-fiction (SF) novels and short stories, radio plays, comic books, computer games, films and TV shows have been encountering alien worlds for much, much longer, using not telescopes or spacecraft but the power of imagination, which has soared way ahead of the science.

It seems strange now, but there once was a time when we didn't even know for sure whether other stars could have planets. Prior to the 1990s the only star that we knew with certainty had planets was the Sun, its eight attendant planets making up the bulk of our solar system: the four inner rocky worlds, Mercury, Venus, Earth and Mars, and the four outer gas and ice giants, Jupiter, Saturn, Uranus and Neptune. (What about Pluto, you ask? It was reclassified as a dwarf planet in 2006, and so belongs in an adjacent category.)

Sure, there were hints that other stars had planets; dusty discs of gas encircling young stars such as the bright summer beacon Vega were inferred to be planetary construction sites. Yet until those first exoplanets were discovered by pioneering astronomers, the only place in which exoplanets existed was science fiction.

And didn't the genre make hay with the concept! The strange new worlds, as Kirk called them, were frequently their own characters, just as important to the fictional narratives as the heroes, villains and starships that took the reader with them on their interstellar adventures. From the desert sands of Arrakis in *Dune* to the twin sunsets of Luke Skywalker's home of Tatooine in *Star Wars*, and the ecological wonder of *Avatar*'s Pandora to the arid, yet no doubt logical, landscape of *Star Trek*'s Vulcan, these and myriad other fictional worlds are indelibly imprinted upon our minds. Before the 1990s science fiction had free rein to come up with whatever wondrous and remarkable worlds it wanted, sometimes hewing closer to the reality of our Earth, other times making leaps of conceptual science or even moving entirely into fantasy.

When astronomers began discovering real exoplanets, things started to change for science fiction. The sands shifted subtly at first, since new discoveries and our understanding of them didn't happen overnight, but with time the burgeoning cottage industry of exoplanet hunting began to overtake the planets of make-believe. Today, exoplanet science is a discovery factory, with billions being spent on the quest for alien worlds and science fiction increasingly having to follow in its stead.

'Now that we know planets are out there, it's different because as a writer you're exploring something that's already defined to some extent scientifically, but it's still very interesting,' the prolific SF writer Stephen Baxter explained when I asked him how real discoveries were changing the ways in which authors were depicting exoplanets in their stories. 'You know the science, and might have some data, so you can use all that as opposed to either deriving it or just imagining it.'

Imagination and speculation still play their part, of course, but now they can be rooted in real science, real discoveries, real worlds. This provides opportunities to compare and contrast. How well do the fictional worlds stack up against the real exoplanets? We'll find out in this book as we explore some of the more notable worlds from science fiction, and to that end I have spoken directly to scientists and science-fiction authors to ask them what SF has got right, what it has wrong and what science can still learn from the fiction.

## The importance of science fiction

It's probably true that most of the public know more about the fictional planets of SF than the real exoplanet discoveries being made by astronomers. Unless you already have a vested interest in the latest research, you're more likely to encounter alien worlds on the silver screen, from your armchair watching television, by immersing yourself in a computer game or in the pages of a novel. And that can mean a big difference between what we, as consumers of SF, think we know about exoplanets and what scientists really do know about them.

'There are a lot of scientific concepts that the general public only ever encounters through science fiction, and exoplanets is one of them,' says Emma Puranen of the University of St Andrews in Scotland. Puranen studies trends in how exoplanets have been depicted in SF, how those trends change in response to scientific discoveries and how SF communicates our understanding of exoplanets to the public.

At its best, science fiction can have a symbiotic relationship with science fact. Science can inform science fiction, and there are many astronomers, some of whom we shall meet in the coming pages, who were inspired to study exoplanets after watching or reading about fictional worlds in SF. One of them is Claire Guimond, a planetary scientist at the University of Cambridge.

'I got into studying planetary science because of the first book in [Frank Herbert's] *Dune* series,' she says. 'I thought it was so cool that Arrakis [the planet in *Dune*] was so well thought out, and it made me think about the science in it a lot. I don't know if it was the only reason why I became a researcher, but I've always been really interested in Arrakis as a planet.'

The effect that SF can have on the public and on scientists themselves through communicating exoplanet science places a lot of responsibility on the shoulders of science-fiction writers, perhaps unfairly – after all, the onus on SF writers is to use exoplanets as a narrative tool to tell a good story, not to contribute to the curriculum. There's a reason it's called 'science' fiction – many of its practitioners and fans love science! And so, SF authors will often pay close attention to the science, going to great lengths to do their research so that their fictional worlds aren't too unbelievable. As we shall see over the coming pages of this book, there's a lot that they do get right, and even with the fictional planets they get wrong, there's still much to learn from their mistakes. 'For better or worse, the public face of exoplanets is science fiction,' says Puranen: 'Astronomers should be paying attention to how science fiction is portraying exoplanets for this reason.'

Puranen has noticed changes between how exoplanets were portrayed in SF before and after the 1990s. The discovery of the first exoplanets was a pivot around which both science and fiction swung, in particular how we started to view these worlds not just as mirrors to our own, but as truly alien and quite possibly inhospitable lands.

For example, prior to the 1990s SF often featured exoplanets that were home to native technological aliens. Think of the home-worlds of the Vulcans, Klingons and Romulans in the various *Star Trek* shows, or Superman's long-destroyed planet of origin, Krypton. Such fictional worlds continue to be written about in stories today, of course, after the discovery of thousands of exo-planets, but Puranen's research suggests that planets that are home to indigenous technological species are not written about as frequently as before.

'I've also noticed that in addition to fewer worlds with native intelligent aliens, there's also a decrease in stories with estab-lished non-native humans,' she says. In other words, there seem to be fewer stories about humans settling on exoplanets, such colonization often being to the detriment of native life. Puranen suggests that this could be because of the growing awareness of the harmfulness of our colonial past, but it's also a reflection of the stark reality that science has cast exoplanets in. When exo-planets only existed in our imagination, it was easy to conjure up fictional societies to inhabit them because storytellers could make those fictional exoplanets as hospitable as they wanted. Now that we know of thousands of exoplanets, none of which, at the time of writing, are known to be habitable like Earth, it's harder to picture worlds upon which societies – either human or alien – could thrive.

It's worth stating again: no other planet like Earth is known. Sure, some of the artist impressions that accompany press releases about exoplanet discoveries depict planets with oceans and clouds, but they're designed to fill in the gaps in our knowl-edge with educated guesses in order to present something to the public. The truth is, while there are some exoplanets that could potentially have a degree of habitability to them, astronomers really only know the barest details about these worlds: their mass, their diameter, the distance at which they orbit their star. Some lie in their star's habitable zone, which is crudely defined as the distance from a star at which a planet with an atmosphere not dissimilar to Earth's is warm enough to support liquid surface water. The logic goes that water is essential for life as we know it, hence why such a distance is called the 'habitable zone', but being in the habitable zone doesn't mean it will have liquid water for certain. Even if it does, that the planet will be habitable – or indeed be inhabited – is not assured. What astronomers have discovered, however, are very many bleak worlds that would be too hot or too cold, or where a significant proportion of their bulk is made from gas. Part of the reason for this is that our techniques for finding exoplanets are currently biased towards finding planets that would be inhospitable, because they are the easiest ones to find. However, it's probably fair to say that SF before the exoplanet discoveries of the 1990s was biased towards imagining worlds that were like something much closer to home. Alas, comfortably habitable worlds like Earth are, so far, in short supply. Instead, at best, we might be looking at habitable niches rather than whole welcoming worlds. Increasingly, more modern SF reflects this; think of the yin-yang world of unbearable heat and deathly cold from Charlie Jane Anders's Locus-award-winning 2019 novel *The City in the Middle of the Night*, or the dark, cloud-smothered moon LV-426 in *Alien* (1979) and *Aliens* (1986) that has to be terraformed to be rendered habitable (although that example actually pre-dates the discovery of exoplanets).

## Distant lands

The planets in science fiction are very close to us – in the pages of the book in our hand, or on the television screen in our living room. Captain Picard's *Enterprise* can zip from one star system to another and back again in the space of one 43-minute episode of *Star Trek: The Next Generation* (1987–94).

In reality, exoplanets are very far away. The closest star to the Sun, which is called Proxima Centauri, lies 4.2 light years away and is home to three exoplanets. Its distance means that its light, travelling at 299,792,458 metres per second, takes 4.2 years to reach us. Effectively, we see it as it was 4.2 years ago – and that's the nearest star system! Other stars and their planets are hundreds, thousands, tens of thousands of light years away, and that's just in our Milky Way galaxy. All those other galaxies in the universe, each one home to billions of stars and countless planets of their own, are millions if not billions of light years away.

The speed of light is the universe's ultimate speed limit; nothing can travel faster. Let's imagine for a moment that we could build a starship, not with magic technology or even the speculative science of warp drives and hyperspace, but with real-life, near-future physics and engineering. In the 1970s the British Interplanetary Society developed Project Daedalus, which was a design for a starship that would propel itself through space using nuclear fusion reactions, capable of reaching 12 per cent of the speed of light, which is equivalent to 35,975 kilometres per second, fast enough to reach Proxima Centauri in 35 years' travel time. More recently, in 2016, scientists using funding from the Breakthrough Foundation began work on Breakthrough Starshot, a project that intends to use energetic laser beams powered by multiple giant fusion reactors on Earth to push a fleet of nano-sized spacecraft to a maximum of 20 per cent of the speed of light (59,958 kilometres per second), capable of reaching Proxima in a little over twenty years. Don't get too excited by these projects just yet – they currently exist on paper alone, and there remain extremely difficult technological and economic challenges to realizing interstellar travel. Even if this was all feasible, it would take spacecraft travelling at these velocities centuries or millennia to reach even more distant exoplanets that lie dozens, hundreds or even thousands of light years away. Of course, science fiction can circumvent this by inventing a warp drive or a hyperspace gate, but if authors want to hew as close to scientific accuracy as they can for their exoplanets, then often the same goes for space travel.

The great distances between us and exoplanets makes detecting them difficult. In our solar system, the planets Mercury, Venus, Mars, Jupiter and Saturn are all very evident because they are close enough and bright enough that we can see them in the night sky with just our eyes. Indeed, they outshine most stars in the night sky. Even Uranus and Neptune, which require a modest amateur telescope to be seen, are both brighter than many of the stars in the sky that we have found to host exoplanets. The stars are just too far away, and if we can barely see them, then seeing their planets is many times more difficult.

At the time of writing, astronomers have only seen about a hundred exoplanets. By see, I mean exactly that: directly photographing them. Astronomers are able to do this on odd occasions by using a device in their telescopes called a coronagraph, which blocks out the glare of a star to improve the contrast by which exoplanets orbiting that star can be seen. Furthermore, the planets that have been imaged tend to be huge, larger than the largest planet in our solar system, Jupiter (which has a diameter of 139,822 kilometres (86,880 mi.)), and several times its mass (which is 1,898 trillion trillion kilograms (4,184 trillion trillion lb)). By necessity, the exoplanets that astronomers have imaged orbit far from their star because otherwise they'd be lost in their star's glare. Also, these directly imaged exoplanets are often relatively young, between a few million and a hundred million years old, meaning that they still contain the heat energy from their birth and thus shine brighter in thermal infrared light. Even taking all this into account, we only see these planets as pinpricks of light, a cluster of pixels on an imaging sensor. While this might sound unimpressive when compared with the imagery of exoplanets in *Star Trek*, for example, what astronomers have been able to image has been profound. One of the best examples is the HR 8799 planetary system, which consists of four giant worlds orbiting a star 133 light years away. Using a coronagraph on one of the 10-metre (33 ft) telescopes at the W. M. Keck Observatory on the peak of Mauna Kea in Hawai'i, astronomers took enough images of HR 8799 to create a time-lapse video showing the planets moving on their orbits around the star like clockwork, following the natural laws of planetary orbits and gravitation laid out by Johannes Kepler and Isaac Newton all those hundreds of years ago. Watching the video is spellbinding.

Speaking of Kepler – nearly four hundred years after the early astronomer found fame, in 2009 a mission to seek out and discover exoplanets was launched bearing his name. From its perch trailing after Earth in its orbit around the Sun, Kepler the space mission has proved to be our most successful exoplanet-hunting project, with 3,326 confirmed exoplanet discoveries as of November 2024. It has found planets by watching for 'transits', which involve the passage of a planet in front of its star, resulting in the planet blocking a small fraction of the star's light. It really is just a small fraction: a planet the size of Jupiter will block about 1 per cent of a Sun-like star's light; an Earth-sized world, just 0.01 per cent. Based on the amount of light blocked, astronomers can calculate the diameter of a planet, and based on how regularly we see that planet transiting, we can calculate the size of its orbit and the length of its 'year'. Alas, we can only see transits for planetary systems with a particular alignment, when their orbital plane is edge-on to our line of sight. There are lots of planets that we cannot see transiting because the alignment is not there.

We find the non-transiting ones through the way they cause their parent star to appear to 'wobble'. Though they are much smaller than stars, planets do still possess significant gravity that can influence their parent star. The balancing point between the star's gravity and an orbiting planet's gravity is called their centre of mass. Since the star's gravity is much stronger, the centre of mass will be closer to the star and more often than not actually inside the star but crucially not at the star's centre. Instead it's shifted a bit off-centre, and by how much depends on the relative masses of the star and the planet, and the distance between them. To envisage this, think of a see-saw with one very large mass on one side (the star) and one comparatively light mass on the other side (the planet). To get the see-saw to balance – in other words, for the centre of mass to be on the pivot – the large mass has to be very close to the pivot, practically on top of it. The star and the planet both orbit this centre of mass, and so the star appears to wobble as its rotational axis shifts around this point. We can't directly see the star wobbling because the motion is very slight – just metres per second. However, even a slow wobble like this can create a Doppler shift. You may have heard of this phenomenon – it's the Doppler shift that causes the pitch of a siren on an emergency vehicle to change as it speeds past you. This happens because the wavelength of the soundwaves is becoming compressed and then stretched again as the vehicle passes. In much the same way, wavelengths of light can become slightly compressed if a star is moving towards us, and slightly stretched if a star is moving away. We can detect this shift in wavelength as a star with an exoplanet wobbles around its centre of mass, and from the amount of shifting and its periodicity, astronomers can determine the mass of the planet orbiting the star and the planet's orbital period, from which the planet's distance from the star can be inferred. More than a thousand planets have now been detected by this method, which is referred to as the 'radial velocity' technique because it is measuring the movement of the star towards and away from us (that is, radially) as it wobbles. If we are lucky enough to be able to measure the radial velocity and see a transit for the same planet, then astronomers can combine the calculated mass and radius of the orbiting planet to determine its density and make a stab at whether it is made from rock, ice/water, gas or some mixture of the three.

## The long legacy of imagining worlds

The grand total of known exoplanets, as of November 2024, stands at more than 5,787 worlds, with several thousand more potential planets still awaiting confirmation. Five thousand, seven hundred and eighty seven alien planets, at least, all ripe for exploration. How many of them are like the worlds presented in SF? How many are considered even stranger than fiction? What unexplored potential do they hold? They represent a fertile landscape across which the imagination can wander, a veritable sandpit of new environments for SF writers to play in.

And those 5,787 exoplanets are merely the tip of the tip of the iceberg. When we look more closely at the statistics, the implications are staggering.

Jessie Christiansen is an astronomer at the California Institute of Technology (Caltech), but she spent her youth under the dark sky of rural Australia, where she fell in love with the stars. Now she's able to experience the thrill of discovering exoplanets, having worked on the Kepler mission, where she describes her role as being 'like Slartibartfas' (from Douglas Adams's *Hitchhiker's Guide to the Galaxy* (original radio play 1978; novel 1979)), in the sense that she simulated hundreds of thousands of light curves of fake planets and input them into the Kepler data to see if their light curves matched any of the dips of light that Kepler was detecting. If they did, there was a good chance that it was a real exoplanet.

'Kepler surveyed 200,000 stars and found about 8,000 candidates around those stars,' she says. Astronomers have to take into account the fact that we can only see transits of planets that happen to be aligned with our line of sight, and that Kepler had limits to its sensitivity. By simulating all those fake planets and comparing them to the data, Christiansen was able to show which planets Kepler could detect and which it would miss, making it possible to extrapolate from there and estimate how common exoplanets are: 'By correcting for the geometry and the detectability of the signal, you turn 8,000 candidate exoplanets into, "oh my god, every star has a planet".'

There are estimated to be at least 100 billion stars in the Milky Way alone, and the Milky Way is just one galaxy among hundreds of billions, if not trillions, in the universe. The numbers are just staggering. Yet we know so little about exoplanets.

This is the dichotomy between the amazing worlds of science fiction and science fact. In SF we often encounter planets with land and sea, vegetation and alien life, climate and habitability, with towering mountains, vast oceans and unending deserts. We get to know them in high resolution and are made to feel like they are real, like we could visit them. Back in reality, we stand at the bleeding edge of our ability to study exoplanets, hoping that NASA's James Webb Space Telescope (JWST), which launched on Christmas Day 2021, will tell us whether any of the relatively nearby rocky exoplanets even have an atmosphere, or that it will sniff out potential 'biosignatures' – the presence of oxygen, water or methane, which could perhaps be attributed to life. It's all very 'low-res' in comparison to the computer-generated vistas of the latest cinematic blockbuster.

'I think this is where science fiction still has a big advantage,' says Puranen. 'Science fiction can provide images of these worlds, as opposed to abstract concepts based on a cluster of pixels.'

Science fiction has been picturing those worlds for a long, long time. Giordano Bruno, a sixteenth-century Italian philosopher who was burned at the stake for his heretical beliefs, spoke in 1584 about the plurality of worlds. The Dutch spectacle-maker Hans Lippershey's invention of the telescope in 1608 and subsequent revolutionary observations by the likes of Galileo, Simon Marius and Thomas Harriot paved the way for the Copernican Revolution – the tenet that Earth is not the centre of the universe and that the planets of the solar system, including Earth, orbit the Sun – and the implication that other stars besides the Sun might also have planets.

Perhaps one of the first science-fiction stories, if not the very first, to feature exoplanets dates all the way back to 1666 and Margaret Cavendish, the Duchess of Newcastle-upon-Tyne. Born in 1623, Cavendish was married to a Royalist commander in the English Civil War and had a keen interest in science and philosophy, becoming the first woman to attend meetings of the Royal Society in London and engage in debate with its male members. In her quasi-SF story *The Blazing World*, the main character travels to another planet orbiting another star by way, somehow, of Earth's North Pole. 'Cavendish seems to be postulating that our Sun might actually be in a binary system, with a companion star we cannot see because our Sun is too bright, or blocks it from view,' says Puranen.

One of the next notable appearances of exoplanets in fiction came in Voltaire's *Micromégas*, published in 1752, in which the eponymous main character comes from a planet orbiting Sirius, also known as the Dog Star, which is the brightest star in the night sky and located just 8.8 light years away (to date, no planets have been identified around Sirius). It was around this time that the natural philosopher Immanuel Kant proposed his 'nebular hypothesis' to explain how the planets of our solar system, and other planetary systems, may have formed. The model was developed further by the French mathematician Pierre-Simon Laplace in 1796.

Remarkably, Kant and Laplace were more or less correct with their hypothesis, which describes how the Sun, planets and the multitude of asteroids and comets that also inhabit the solar system all formed from a collapsing, spinning cloud of gas and dust that today we refer to as the protoplanetary nebula. Modern telescopes can actually see such clouds and the stars forming within them. However, at the turn of the twentieth century, there was an apparent problem with the nebular hypothesis.

A massive, rotating cloud of gas and dust, collapsing under its own gravity, contains lots of angular momentum. Think of angular momentum as being the momentum (calculated by the mass × its velocity) of something spinning around. Momentum, like energy, must be conserved in a system – it's one of the fundamental laws of nature. As the cloud collapses, the burgeoning Sun at the centre should start spinning faster and faster as all the angular momentum is concentrated there. Think of an ice skater spinning faster with their arms tucked in rather than outstretched. The problem is that today the Sun's angular momentum is pretty low, since it leisurely rotates once every 27 days. Where did all the angular momentum go?

Late Victorian astronomers noticed this, and it put them off the nebular hypothesis. Instead they latched on to an alternative, which was the idea that the gravity of a passing star had ripped a significant amount of material from the Sun, and it was this material that had become the building blocks of the planets. However, the great distances between the stars as they move on their respective paths through our galaxy means that such close encounters would be rare; therefore, if this theory were true, planets would also be rare, and our solar system would be a fluke. The possibility that there could be planets around other stars was brought seriously into question.

The science-fiction writers of the time reacted to this by mostly keeping their stories set within the confines of our solar system. Visiting planets such as Mars, as featured in H. G. Wells's *The War of the Worlds* (1898) and Edgar Rice Burroughs's stories of John Carter (published between 1912 and 1964), was adventure enough. The stellar encounter hypothesis posited that the first planets to form from the material ripped from the Sun would be those furthest out. So in this scenario, Neptune would be the oldest planet, then Uranus, Saturn, Jupiter, Mars and then Earth. This was reflected in some of the science fiction of the time; the Martians in *The War of the Worlds* were more technologically advanced than the humans of the Victorian era, with their Penny

Farthings and dreadnoughts, because their planet was supposed to be older than ours – a kind of logic that ignored the vagaries and uneven course of evolution by natural election. The official sequel to *The War of the Worlds*, Stephen Baxter's 2017 novel *The Massacre of Mankind*, used this idea of the outer planets and their inhabitants being older as a plot point, with a plea to the hierarchy of worlds ultimately bringing the second Martian invasion to an end.

In light of this, science fiction from this period that chose to explicitly visit exoplanets was a serious departure from the scientific consensus of the time. One example is *The Skylark of Space* by E. E. 'Doc' Smith, who was also a food technologist famous for finding a way to get powdered sugar to stick to dough. So the next time you chomp down on a jam doughnut, think of Smith. Despite the ubiquitous popularity of doughnuts, however, Smith is even better known as an author who trailblazed the 'space opera' genre (for those unfamiliar with the term, think spaceships, swashbuckling and interstellar travel), even if his writing is somewhat dated in both style and attitudes today. *The Skylark of Space* was written between 1915 and 1921 and finally published in 1928, the first of a series of novels that describe the escapades of Dick Seaton and his spaceship, the eponymous *Skylark*.

In the story Smith takes the reader to a variety of exoplanets. Some of them resemble Earth, but one nameless world stands out for being located within a small, tight cluster of seventeen stars, with an ocean of copper sulphate on its surface. The properties of the planet are puzzling to say the least: air pressure is twice that at Earth's surface, but with just two-fifths of our gravity. How a small world with such low mass, and therefore low gravity, can hold on to such a dense atmosphere (not least have an ocean of pure copper sulphate, other than as a convenience for the hero, who needs copper to power his spaceship) is something of a paradox, and this is exactly the kind of question we'll be asking in the coming pages.

The stellar encounter theory of planet formation gradually fell into disfavour, thanks partly to astronomical observations that revealed many young stars to be surrounded by discs of gas and dust that are forming planets, just as the nebular hypothesis predicted. The angular momentum problem still persists somewhat, although scientists now think that stars shed their angular momentum by transferring it to the planet-forming disc around them. The remaining unanswered questions relate to the actual planet-building process: how rocky planets are assembled through a process of runaway accretion, and whether gas giants form from the top down, condensing directly out of the disc, or from the bottom up, growing a massive, rocky core before their gravity sweeps up plentiful amounts of gas.

The space opera genre became the mechanism for SF authors to explore the other worlds of the Milky Way. Readers went on adventures with their heroes, fought battles in space, witnessed empires rise and fall, civilizations grow and evolve, met weird and wonderful alien life and, in the deeper works of the genre, examined culture and society, life, death and rebirth, all through the lens of space adventure. In doing so, they visited all the strange new worlds that our imagination had to offer, filling in the gaps left by the science of the time and the limited power of the telescopes of yesteryear. And so, Isaac Asimov's *Foundation* grew on a world called Trantor; Robert Heinlein's *Starship Troopers* fought the bugs on a planet called Klendathu; Paul Atreides overthrew an empire from the desert world of Arrakis in Frank Herbert's *Dune*; Ursula K. Le Guin asked difficult questions of gender, nationalism and first contact on the frozen planet of Gethen in *The Left Hand of Darkness*; and on film Luke Skywalker took his first steps into a larger world while watching the double sunset on Tatooine in *Star Wars*. Exoplanets have been the hallowed turf on which our favourite characters have become heroes or villains, on which profound questions about the nature of humanity have been asked, and on which the greatest epics of the genre have played out. They also filled the gaps in our knowledge, at least until the first real exoplanets were discovered, and inspired young people passionate about science to go into a career looking for real life exoplanets to match the lands that they read about or saw at the cinema.

'Science fiction definitely inspired me to go into science in the first place – we're turning science fiction into fact,' says Knicole Colón of NASA's Goddard Space Flight Center, who works as the deputy project scientist for exoplanet science on the mighty JWST. She continues, 'And now I love that we can now inspire authors and screen writers to take real science and turn it into fiction.'

As we progress through the pages that follow, we'll learn more about this symbiotic relationship between science fiction and exoplanet science, how each group inspires the other and how, sometimes, they work together to create fictional planets that have roots in the strange new worlds falling prey to our telescopes. And while most of those real exoplanets are too alien to be hospitable, a few may have what it takes to be called 'home'.

# 2

# HOME AWAY FROM HOME

In the cliffhanger season six finale of *Star Trek: The Next Generation*, the crew beam down onto a planet with a beautiful landscape of giant oaks growing from hillsides and expansive grass meadows. The daylight appears golden, saturated, as though the light from the planet's star is a different hue to our own Sun.

It doesn't fool us though. This is no alien planet. It's just the Californian countryside viewed through a camera filter. In fact, a lot of the planets in the *Star Trek* universe look like California, the limitations of filming science-fiction shows in Hollywood. Sometimes productions make a bit of extra effort – for the scenes in the forests of Endor in *Return of the Jedi* (1983), which were filmed in a redwood forest in Del Norte County in California, set dressers brought in non-native plants to make the forest look a little more exotic, but they still looked like earthly plants.

If you're a fan of *Star Trek* or *Star Wars*, and that's your only exposure to the concept of exoplanets, then you might be forgiven for thinking that other planets are just like Earth, down to having literal carpets of grass. Even many SF novels, which aren't limited by budgets or filming on Earth with human actors dressed as aliens, often feature very Earth-like planets with – if not exact replicas, then analogues of – trees and grass, fruit and flowers, birds and animals and insects, and a pleasingly breathable atmosphere. Of course, these fictional planets are a means to an end, which is the story that is being told. If you want your heroes to explore a strange new world, it can sometimes be convenient not to make it too strange, or else it will be too hostile a place for your heroes to visit.

In reality the universe is not beholden to us to build planets that we'll find comfortable. Earth inhabits a sweet spot of clement temperatures, ample liquid surface water and an oxygen-rich atmosphere. As we shall see throughout this chapter, and indeed this book, Earth is like a pin balancing on its point, and events can conspire to tip it over one way or the other into relative uninhabitability, as has happened in the past and will happen again in the future.

We'll explore some of those extremes – dry desert worlds, frozen snowball planets, global oceans and the fictional planets that attempt to portray them – in the coming chapters. And given how much of a Goldilocks planet Earth seems to be – it has 'just right' levels of temperature, water, land, oxygen, gravity, plate tectonics (and their resulting natural disasters, volcanoes and earthquakes), a stable Sun, an infrequent asteroid and comet impact rate, and on and on – it might be reasonable to suggest that planets that are habitable in the way that Earth is habitable might not be common. (I'm being careful with how I word things here – planets substantially different to Earth could still be habitable in their own way, but they would not be like Earth.)

Before we go any further, I'd like to pause so that I can remind the reader that although astronomers have discovered more than 5,787 exoplanets at the time of writing, none of them are known to be like Earth. None of them are known to be habitable. None of them are known to be wet with liquid water flowing across their surface. And none of them are known to have life. That doesn't mean that such worlds don't exist; it doesn't even mean that we haven't already found such a world, it just means we don't yet have the ability to tell whether we have or not.

Sometimes this deficiency in our exploration of exoplanets is glossed over. Whenever a notable exoplanet discovery is made, it's undoubtedly proclaimed in the news as the detection of a new 'Earth-like' planet. What they really mean is 'Earth-sized' and possibly in their star's habitable zone – we cannot yet tell whether it is really like Earth, with oceans and continents and vegetation and life. What we can definitely discern based on the techniques described in Chapter One is an exoplanet's physical characteristics: its mass and radius, its orbital period, and hence distance from its star, and how much energy it therefore receives from its star. If we know both its mass and its radius, we can calculate its density and state with confidence whether it is gaseous, rocky or some water–gas–rock hybrid.

We can then input these data into the Earth Similarity Index (ESI), which was devised by a group of planetary scientists in 2011. The index's main cheerleader is Abel Méndez, a planetary scientist and astrobiologist from the University of Puerto Rico at Arecibo, where he is the director of the Planetary Habitability Laboratory.

'The Earth Similarity Index is a multi-parameter method to calculate a similarity to Earth based on a set of properties,' says Méndez. So we can take those aforementioned physical properties – mass, radius and so on – and compare them to our Earth. The most Earth-like planet in this regard is a rocky world orbiting a little-known red dwarf called Teegarden's Star (a small, cool stellar body – see Chapter Six for more), which is just 12.5 light years away – virtually in our Sun's back garden. There are actually two planets there, both orbiting within the nominal habitable zone of their star, but the innermost planet, b, is considered the most promising. If Earth has an ESI rating of 1, then planet b around Teegarden's Star has a rating of 0.95. However, all that tells us is that it is about the same size and mass as Earth, and that it receives almost the same amount of energy from its star as Earth does. It doesn't tell us that it's habitable; it doesn't tell us that it's like Earth. Méndez points out that Venus, for example – scorching, suffocating, toxic Venus, with the hottest surface temperature in the solar system reaching 475 degrees Celsius – also has a similar ESI rating to Earth. 'That makes sense if we can imagine that we don't know anything about the atmosphere of Venus or its surface conditions,' says Méndez. 'We even call Venus our sister planet!'

Venus actually resides on the inner edge of our solar system's habitable zone. Its diameter is 12,104 kilometres (7,520 mi.), just 638 kilometres (396 mi.) smaller than Earth, and with 82 per cent of the mass of our planet. Yet all that choking carbon dioxide in the atmosphere has turned Venus almost – but not quite, as we'll come to later – totally inhospitable, since it has generated a runaway greenhouse effect. 'If Venus had less atmosphere it would be at a nice temperature!' adds Méndez.

## The search for biosignatures

Having been exposed to countless habitable planets in science fiction, it is frustrating, for astronomers and SF fans alike, to be hitting this brick wall where we cannot see past an exoplanet's basic bulk properties, but that time is coming. Already astronomers have performed indirect observations of the atmospheres of giant, gas-rich exoplanets that are close to their star – so-called 'hot Jupiters' – which constitute most of the first exoplanets to be discovered simply because their size and proximity to their star make them stand out. Today, NASA's James Webb Space

Telescope (JWST) is similarly able to perform some limited atmospheric studies of the closest Earth-sized exoplanets – those around Teegarden's Star, the septuplet of worlds orbiting TRAPPIST-1 and a few others. Some preliminary results have already come in, and by the time you read this book JWST may have completed this atmospheric reconnaissance. Like light being split into a rainbow of colours by a prism, JWST is able to perform something called spectroscopy, whereby light from an object is divided into its individual wavelengths. Astronomers can then look for bright spectral emission lines or dark absorption bands related to specific atoms and molecules.

JWST's spectroscopic observations will be able to tell us about some of the more prominent gases in the atmosphere of these planets – if they have an atmosphere at all – and possibly, if we're really lucky, some hints of biosignatures, if indeed any are present. We alluded to 'biosignatures' in the previous chapter, which in the context of this discussion are gases in a planet's atmosphere that could be connected to the processes of life. These gases are said to be in disequilibrium, in that they are being produced and replenished in quantities that cannot be explained by more mundane geological or atmospheric chemical reactions. For example, methane and molecular oxygen in Earth's atmosphere are in disequilibrium because their abundance is the result of waste gases produced by biological activity (think of cows farting, or photosynthesis). Finding atmospheric methane doesn't necessarily mean there is life, since it can also be produced by geochemical processes, but finding an abundance, possibly with other biosignature gases too, could be a big clue that there is some form of biological activity.

Finding something like that in the atmosphere of an exoplanet would be tremendously exciting, but for the really good stuff we need to wait until after the 2040s, when NASA, the European Space Agency and others aim to launch their next-generation space telescopes. These telescopes will come equipped with the ability to directly image some of these smaller planets, rather than just relying on indirect transits or Doppler shift measurements. A directly imaged exoplanet won't appear as anything more than a pinprick of light, but even from that tiny dot a lot of information will be gleaned. For example, land and sea reflect light differently, so as the planet rotates we will be able to detect this change depending on whether we are looking at continents or oceans. We'll also be able to directly measure the planet's surface temperature, and really discover whether it's more like Earth, or more like Venus.

However, 'Direct imaging is much harder than the spectroscopy JWST is doing,' says Robin Wordsworth, a planetary scientist from Harvard University. As we described in the previous chapter, direct imaging requires the use of an instrument called a coronagraph to block the intense glare of the host star so that the feeble light of the planet next door can be seen. 'But that's next generation technology,' adds Wordsworth: 'We can do it now for some planets that are young, massive and quite far from their star, but to look at an Earth-sized planet around a Sun-like star, we're talking about a mission that's still in the planning phases.'

Perhaps those future telescopes will finally be able to tell us whether Earth-like planets are regular occurrences around other stars or not. It would mean a lot to us if they were. Hunting for Earth-like planets has become a kind of holy grail for many astronomers, because finding another world like Earth would provide some cosmic context for our own existence. It would also fuel dreams of being able to travel among the stars and settle on other worlds that are comfortably familiar and suitable for life as we know it, and perhaps we may even meet extraterrestrial life that on some level we can recognize.

Until then it remains the province of science fiction to describe what habitable exoplanets might be like. In the previous chapter I wrote about how there was a demarcation in SF between stories written before the discovery of exoplanets and stories written afterwards. There's another sub-demarcation here: we are still in the era before habitable exoplanets and extraterrestrial life have been discovered. That affords SF writers some latitude in how they portray such things. Since we have a really good example of a habitable planet right here on Earth, it is unsurprising that SF draws from this well so often.

## Alien plant life

*Star Trek* has always treated Earth-like exoplanets as mundanely common. This isn't necessarily a slight against *Star Trek* – how awesome would it be if that were true! The example from *The Next Generation* at the beginning of this chapter is but one of dozens, if not hundreds, of Earth-like planets visited during the various *Star Trek* shows – so many that Emma Puranen has purposely limited the number of planets from *Star Trek* in her fictional planet database simply so that they don't overwhelm it. 'And wow, a lot of the *Star Trek* planets are pretty boring and uncreative clones of Earth,' she says, being careful to stress that although *Star Trek*'s planets generally don't impress her, she is a fan of everything else *Trek*.

This is partly a symptom of the limited budgets on TV shows, whereas authors of SF novels only need to worry about limited imagination. But let's not get too snobbish about SF literature – the tendency to rely on Earth-like worlds is common among written SF as well, particularly in the space-opera genre, where adventure is often the operative word rather than scientific accuracy, and in Golden Age science fiction dating from before even the planets in our own solar system had been characterized.

One of the greatest space-opera sagas in the modern era is Peter F. Hamilton's *Night's Dawn* trilogy, published in the late 1990s. So as not to spoil the story, I won't go into details other than to say it involves a star-spanning Confederation of worlds settled by humans in the year 2611. Much of the action in the first book in the trilogy, *The Reality Dysfunction* (1996), is set on the planet of Lalonde, which Hamilton describes as a 'terracompatible' (that is, Earth-like) world orbiting a nondescript Sun-like star located 319 light years from Earth. Thank goodness for Hamilton's *Confederation Handbook* (2000), a companion piece to the trilogy that describes in great detail the worlds of the Confederation, including Lalonde. It tells us that Lalonde orbits at a distance of 132 million kilometres from its star, from which we can draw our first comparison with Earth: orbital distance. We orbit the Sun at a mean distance of 149.6 million kilometres, so a little further out than Lalonde, whereas Venus orbits at 108 million kilometres, a little closer in, so in our solar system Lalonde would be firmly within the habitable zone. This is countered somewhat by Lalonde's star being a little smaller (by about 5 per cent) and a little cooler than our Sun, radiating with a temperature of 5,276 degrees Celsius compared to the Sun's 5,500 degrees Celsius. (I should point out, for the nitpickers among you, that Hamilton contradicts himself, describing Lalonde's star as a slightly hotter F-type in the novel, but as a cooler G7-type in his *Confederation Handbook*; I've chosen to go with the latter for the purposes of this discussion.) On balance, Lalonde's tropical climate, with equatorial regions inhospitable to humans, doesn't sound too unreasonable.

More intriguing is Lalonde's vegetation. One character, viewing the planet through the window of a descending spaceplane, describes Lalonde as a world of trees, 'real bloody trees! Millions of them! Trillions of them! A whole bloody world of them!' Indeed, the planet is largely jungle with several types of trees, along with a carpet of 'bushy grass', and the logging of the Mayope tree in particular forms a significant part of the planet's fledgling economy. The grass, the trees, the vines and reeds and fruit, all feel very Earth-like, but how realistic is it?

'Tree-like things have evolved quite a few times,' says one of Hamilton's contemporaries, the SF author Paul McAuley, who also sports a PhD in botany (specifically in plant–animal symbiosis). Across history on Earth, there have been 'mushroom tree-like things, fern tree-like things, cactus tree-like things, tree tree-like things', he says.

Let's sidestep into evolutionary science for a moment. Darwinian evolution is often boiled down to 'survival of the fittest': nature selects genetic traits that give a life form an advantage. It's somewhat random, rather than being by design, and because nature cannot see into the future, it often leads to dead ends. We can see this in one of the greatest paleontological finds in history, the Burgess Shale, a rock formation in the Canadian Rocky Mountains that contains the fossils of bizarre creatures dating back to an event that occurred 505 million years ago, the Cambrian explosion. Despite the name, the Cambrian explosion wasn't a disaster – far from it. Prior to this event life on Earth had been limited to microbes: single-celled amoeba, bacteria, that kind of thing. However, something – and scientists still aren't sure what – prompted evolution to suddenly speed up, and an amazing variety of complex lifeforms 'exploded' into being in a relatively short geological timeframe. Many of the fossils in the Burgess

Shale, a relic of that time, defy classification. Take for example the Wiwaxia, which was like an armoured slug – a kind of elliptically shaped, squared-off 5-centimetre-long (2 in.) blob with armour-plated scales, rows of defensive spines and conical teeth on its underside for eating whatever it was it preyed upon. Even weirder was Hallucigenia, appropriately named. A water-dwelling species, it looked like a worm or an eel, but with fourteen sharp, pointed spines on its back and a similar number of tentacled limbs on its underside, presumably for walking and foraging on the seabed.

The Cambrian explosion was a vital period in the history of life on Earth because it introduced complex eukaryotic (multicelled) life-forms to the planet for the first time. The distant ancestors of all animal life on Earth can be traced back to the Cambrian explosion, but there also seems to have been a lot of wastage. Though it's still debated, many of these life forms did not seem to have long lineages – we struggle to find their descendants today. Evolution, given its surprise shot in the arm during the Cambrian explosion, was trying out many new things. Some things worked, and either survived or were reinvented further down the line because they nicely filled an evolutionary niche. For example, eyes have evolved independently hundreds, perhaps thousands, of times throughout Earth's history – from octopus's eyes to mammalian eyes like our own. The zoologist Andrew Parker, in his book *In the Blink of an Eye: How Vision Sparked the Big Bang of Evolution* (2004), even suggested that developing eyes was the driving force behind the Cambrian explosion, since it raised the stakes between predator and prey, and between survival and death.

We call the repeated evolutionary invention of eyes – and trees – convergent evolution, in the sense that evolution converges on similar solutions because they work well. Climbing high to reach precious sunlight obviously provides a survival advantage for photosynthesizing plants, hence trees. So, by that logic, we may expect to see more tree-like things on other Earth-like exoplanets. They won't exactly be trees as we know them, but they'll be roughly analogous in the sense that they'll fill the same evolutionary niche.

Grass, on the other hand, is a much more recent invention by nature. Its origin has been placed somewhere between 55 and 66 million years ago, so basically just after the age of dinosaurs. Including grass is an easy mistake that prehistoric dinosaur films make (a recent offender being the SF flick *65* (2023), starring Adam Driver, about a human-looking alien who crash-lands on Earth at the end of the dinosaur-dominated Cretaceous period, of which it happens to get the date wrong). It is perhaps, too, a mistake that SF authors looking to describe the alien vegetation on exoplanets make.

'You can look back [in time to before there was grass] and see that there were giant mushroom plants, and before then there were lichens and some weird horsetail stuff, but there was no grass,' says McAuley. 'Grass is a very human thing, our hominid ancestors were shaped by grasslands as they went out into the savannah as the forests were shrinking because Earth was undergoing a dry period.'

Given that it is a relatively new thing compared to other types of plant, and has, to our knowledge, only been invented once by evolution, grass is probably an example of what we describe as contingent, rather than convergent, evolution. Unlike tree-like plants, grass seems to have evolved once and only once. There are different species of grass, of course, but they all evolved from the same grassy ancestor. So we should not expect to see grass on other planets.

Needless to say, to some degree we're talking about appearances. A cactus might look a bit like a tree, but it is not an actual tree (it lacks the woody trunk). It's merely a tree analogue, something that resembles a tree and fills a similar niche. Sometimes science fiction will delve beyond appearances to find these interesting niche-filling differences. In *The Expanse*, a series of novels by James S. A. Corey (a pseudonym for two authors, Ty Franck and Daniel Abraham) published between 2011 and 2022, a fascist human empire sets up home on a distant world called Laconia, accessible only via the 'ring gates' that provide passage through some kind of hyperspace. On the face of it, Laconia is very Earth-like, with trees and plants, seasons and weather, an oxygen–nitrogen atmosphere and a water cycle. However, deep down in Laconian life's molecular bonds, there is a huge difference between it and life on Earth.

All life on Earth is left-handed. What we mean by this is that the amino acids that are components of the proteins found in RNA and DNA in our cells lack rotational symmetry. This is analogous to how, no matter how much someone's left hand is turned, it can never be superimposed on their right hand. They are equal but opposite, and we call this property chirality. Amino acids are chiral too, and all the amino acids in our cells are 'left-handed'. However, all the sugars used in our cells – as well as ribosome, which links amino acids into RNA chains – are 'right-handed'. Why biology should prefer this organization of left- and right-handedness is unclear, especially given that both left- and right-handed amino acids are present in non-biological circumstances on Earth – it's not as if nature didn't have the raw materials.

Amino acids have been found in meteorites that have fallen to ground from space, having been blasted off asteroids. They have even been identified in an asteroid called Ryugu, by the Japanese mission Hayabusa2, which brought samples of the asteroid back to Earth in 2020. Asteroids are the residual building blocks of the planets, and the presence of amino acids within them suggests that amino acids were brought to Earth in impacts, rather than having originated here. There is even evidence that amino acids could exist within the vast clouds of molecular gas that are the birthplace of stars and planetary systems – while amino acids themselves have not been detected in these molecular clouds, their biological predecessors, namely organic compounds called amines, have been spotted among the stellar embryos. Amino acids might not have been brought only to Earth – they might have been brought everywhere, forming part of the biological machinery of alien life on a multitude of worlds. The question is, what would its handedness be? A sure sign of alien life would be proteins that exclusively use right-handed amino acids, as opposed to our left-handed variety, and this is exactly the case for life on Laconia.

'Laconia is very Earth-like. It's forested, people can breathe there, but the life has mirror chirality so that the handedness of the amino acids and sugars is flipped, meaning people don't get any sustenance from eating the food on the planet,' says Emma Puranen. A similar scenario plays out in Becky Chambers's 2019 novella *To Be Taught, if Fortunate*, where astronauts exploring an exoplanetary system discover primitive life with right-handed amino acids and left-handed sugars. To both Chambers and her lead character and narrator, Ariadne O'Neill, this is a profound moment, because it means that life in the universe need not have begun exactly the same way that it did on Earth. Life will make use of whatever tools it has at its disposal, and those tools are brought from space by meteorites, which means that the entire galaxy is filled with the raw material for life; it has been seeded everywhere.

In *The Expanse*, it is presented as a small, inconsequential detail lacking the profundity of Chambers's story, but nevertheless, it adds a degree of 'alienness' while introducing a scientific concept that will be new to many readers. 'I appreciate worlds that do interesting or unexpected things like that,' says Puranen.

## Niche worlds

Speaking of niches, not all of them are evolutionary. Sometimes they are geographic, or climatic, a little piece of temperate Earth amid a hostile larger environment, like an oasis in the desert or a fragile flower peeking out from the arctic ice. Seeing as we've already mentioned it in this chapter, let's start with a real example: Venus. We know that its surface is truly inhospitable. It's not just the insane temperature, it's the crushing pressure that can crumple spacecraft under 92 bars of dense, noxious atmosphere. The first missions to land on Venus, the Soviet Venera probes, lasted at best two hours before succumbing to the conditions, but it's hoped that in the future new lander missions will survive for longer, with hardened armatures to resist the temperature and pressure combo.

However, high above the surface, at altitudes of between 51 and 62 kilometres (167,000 and 203,412 ft), things are much more Earth-like. The temperature ranges between 65 and −20 degrees Celsius, with the most comfortable range from 37 degrees Celsius at 51 kilometres to 20 degrees Celsius at 62 kilometres, as measured by the European Space Agency's Venus Express orbiter mission. A bonus is that at this altitude the atmospheric pressure is akin to the surface pressure on Earth. This 'habitable height' isn't without its difficulties – there are still clouds of toxic sulphur dioxide floating about and a near total lack of water vapour, but if there's life on Venus, this is where we would find it. For what it's worth, the controversial discovery of the gas phosphine, which on Earth is only produced by biological processes, was made at this habitable altitude on Venus. However, the discovery of phosphine remains controversial (although more scientists are coming around to the idea), so we should file it away as 'unknown' for now.

On Venus you have to travel upwards to find a temperature and pressure suitable for life, but we might imagine on some worlds the air is so thin that we need to go down instead. In Alastair Reynolds's 2001 novel *Chasm City*, the eponymous city is, as the name suggests, found deep within a chasm on a fairly barren planet called Yellowstone, in orbit around the real star epsilon Eridani, which is just a short 10.5 light years from Earth. Epsilon Eridani is a popular star in science fiction, possibly because of its relative proximity – in the TV series *Babylon 5* (1993–8), the eponymous space station orbits the third planet around the star, a world called Epsilon III that's just as barren as Yellowstone, and which also hides technological life deep within a fissure.

Anyway, the chasm on Yellowstone retains warmth better than the rest of the planet, and outgassing from vents provides and replenishes a denser, breathable atmosphere. It's very much a liveable niche on an unliveable planet, but Reynolds says that the inspiration for Chasm City comes not from Venus or Mars, or from any other scientific discoveries, but from the *Known Space* series of stories written by Larry Niven (the most famous of which is the award-winning novel *Ringworld* (1970), set on an artificial world that encircles its star).

Early in the chronology of the *Known Space* universe, humans sent out robotic survey craft to find Earth-like exoplanets that humans could settle on. Alas, an error in the robots' programming meant that they would report a positive discovery even if just one tiny part of the planet was habitable.

'You'd probably make your programming a bit cleverer than that,' quips Reynolds, but he goes on to say that 'this was Niven's fun way of coming up with all these human colonies on planets that were a bit weird in terms of the standard at the time.'

The *Known Space* stories were also a favourite of Paul McAuley's, who highlights how 'Niven had this whole variety of worlds, and these worlds all had Earth-like bits, but none were particularly Earth-like as a whole.'

One of these worlds, named Plateau by its settlers in Niven's 1968 novel *A Gift from Earth*, is a Venus-like world orbiting another nearby star, tau Ceti, which is 11.9 light years away. In real life, tau Ceti has four confirmed exoplanets ranging in mass from 1.75 to 4 times the mass of Earth ($5.97 \times 10^{24}$ kilograms, or $13 \times 10^{24}$ lb), meaning that they are either oversized rocky planets called 'super-Earths', or possibly 'mini-Neptunes' that have migrated in from the cold. These four worlds might not be alone; there's tentative evidence for a further four planets within the same mass range, but to confirm whether they are real or not, astronomers must gather more data.

But back to the fiction: like Venus, the vast majority of Plateau is uninhabitable, except for its giant roughly 600-kilometre-wide (373 mi.) plateau (hence the name), christened Mount Lookitthat by the settlers. Mount Lookitthat rises high up into the atmosphere, reaching the region where conditions are habitable, somewhat similarly to Venus.

'There's just this one mountain sticking out with a habitable point 80 kilometres [50 mi.] above the surface, and that's where the colony is,' says Reynolds. The *Known Space* planets inspired the teenage Reynolds in the early 1980s to begin developing

*Chasm City* and other stories set in the same universe (particularly his debut novel, *Revelation Space* (2000)).

'Niven played with the idea of lots of little niches,' says Reynolds. 'As a teenager I was so in thrall of that, so with Yellowstone I just tried to make a Larry Niven-type planet, basically. I thought, what if the only survivable location on this planet is inside a chasm that's a bit warmer and less toxic than the rest of the planet?'

A similar niche is found in Peter F. Hamilton's 2018 novel *Salvation*, the first in a trilogy about an alien invasion. A sub-story that forms part of the background for one of the main characters features a penal colony in a 7-kilometre-deep (4½ mi.) chasm on a barren planet called Zagreus that is only reached by wormhole. In the story, Zagreus orbits alpha Centauri A, which is the triple-star system that Proxima Centauri is part of. So far, no planets have been discovered orbiting alpha Centauri A or B, but it would be surprising if there are none. Zagreus orbits three times further from its star than Earth is from the Sun, so beyond the habitable zone. It's a rocky planet, slightly bigger than Earth but with an atmosphere as thin as Mars's. It's cold and hostile; one character mistakes it for Antarctica at night. The only habitable region on the entire planet is inside this deep chasm, where lakes fill volcanic calderas and gases vent into the air – the last pocket of air on the planet.

Sometimes SF authors explore these niches not by transporting their characters to exoplanets but by exploring other Earths in parallel universes. This gives authors a chance to keep Earth mostly the same, but then experiment by tweaking this or that. This is the key plot device in Stephen Baxter and Terry Pratchett's five-novel *Long Earth* series (2012–16), where people can jump to myriad neighbouring parallel Earths with the aid of a 'stepper'. The more one 'steps', the more the changes add up.

'It's delightfully imaginative,' says Adiv Paradise, a planetary scientist at the University of Toronto in Canada. 'It does fall a little bit into the trap of this version of Earth is one environment, that version is another environment and so on, but it does a great job of exploring how small changes add up over time, and how different Earth can become through these changes.'

Baxter himself highlights another story set in a parallel universe that throws up a 'what if?', specifically what if Mars was larger and more Earth-like? Harry Turtledove's 1990 novel *A World of Difference* explores this alternative reality, where Mars is a super-Earth named Minerva populated by a non-technological intelligent species.

'It's an alternate history novel,' says Baxter: 'Instead of Mars there's an Earth-like planet, actually bigger than Earth with a thicker atmosphere. The Americans and Russians are racing to try and get to this planet, and then everything changes.'

The story highlights the puzzling fact that our solar system lacks a super-Earth planet. It's puzzling because astronomers are finding super-Earths quite frequently around other stars, so why should our solar system be lacking one? Turtledove's alternative timeline might be hinting at something. Mars is small, just 6,779 kilometres (4,212 mi.) across, which is about half the size of Earth. It could have grown bigger, except that, according to one popular theory, early in the history of our solar system Jupiter migrated from where it formed to closer to the Sun. It was eventually dragged back out by the gravity of the pursuing Saturn, but its dalliance with the inner solar system was enough for gravitational tides emanating from the giant fifth planet in the solar system to clear out much of the planet-forming material from around the embryonic Mars. As such, starved of material from which to grow, the Red Planet ended up being a bit of a runt.

Who knows how massive it could have grown had it been left to its own devices? A more massive Mars would have meant more gravity to hang on to its atmosphere, which has instead leaked into space over billions of years, taking much of the Red Planet's water with it. A more massive Mars with an atmosphere would be warmer, potentially able to support liquid water on its surface for longer, and where there is water there could be life. Instead, Mars's habitability was on a knife edge, and about 3 billion years ago it toppled over into relative uninhabitability as the planet lost its atmosphere and cooled (we'll discuss the effects that Jupiter's migration had on Mars some more in Chapter Four).

As we have previously mentioned, Earth's habitable environment is also balanced on a knife edge. And as we are witnessing at first hand with climate change, it doesn't take much to change those conditions for the worse. The reality is that finding many exoplanets balanced just so on that same knife edge is going to be unlikely. Rather, exoplanets might be missing some of the characteristics that make Earth habitable, or maybe possess some extra ones that Earth doesn't have. It might be that niches are more common than global habitats. In the coming chapters we will explore some of those niches, patches of habitability on worlds that are too hot and dry, or worlds that are too cold, or are drowning in water, or even worlds that only show one face to their star. None are what we would call Earth-like, yet all to varying degrees have been shown, in theory, to be potentially habitable, and have been explored, in depth, in science fiction.

# 3

# LANDS OF SAND

Winds laced with caustic sand whip through the air as the sun, or suns, beat down, and on the horizon a dark cloud lies, a dust storm the size of a continent marauding across the land. For as far as the eye can see, dunes ripple as they recede into the shimmering heat haze. The sky above is clear of clouds, and nowhere can a drop of water be found.

Desert planets are a staple of science fiction. It's the old conceit – take a singular environment from Earth and extend it to cover an entire planet. Some of the most famous fictional worlds are desert planets. Luke Skywalker couldn't wait to get away from his desolate desert homeworld of Tatooine in *Star Wars* (1977), despite its beautiful double sunsets, while one of the best and most famous stories in all of science fiction, Frank Herbert's *Dune* (1965), is about the battle to control a very important, but very inhospitable, desert world called Arrakis.

Despite appearances to the contrary, it turns out that desert worlds might be eminently inhabitable. More so, they may be the most common type of habitable planet in the galaxy, a far cry from the lush, but scarcer, paradise of Earth. Tatooine and Arrakis are fictional, but they are not necessarily unrealistic. In fact, Frank Herbert spent so much time carefully developing the ecology of Arrakis that studies have shown how very similar worlds – well, perhaps without the sandworms – could exist around a star possibly not too far away from us.

This revelation that liveable desert worlds like Arrakis might outnumber Earth-like worlds promises to transform how we think about habitable planets in the universe. Yet for now, the search for another planet just like Earth remains the holy grail of exoplanet research – we just can't seem to get past our Earth-based chauvinism.

We first encountered the concept of the habitable zone in chapters One and Two. It is defined as the distance from a star where temperatures are suitable for liquid water to exist on the surface of a planet. It's sometimes referred to as the Goldilocks zone, for being 'just right' for life, but the more we explore the concept, the more vague the limits of the habitable zone become.

'Just right', in this sense, means having liquid water. Life as we know it cannot exist without water. Biologically, water acts as a solvent for many of life's most valuable molecules, from proteins to DNA strands. Every form of life we have ever encountered on Earth requires at least some water. Some leading theories about how life began on Earth even suggest that the first life developed in deep ocean vents, or tidal pools.

'If there's water there will be life, I'm pretty convinced of it,' says Dale DeNardo, a professor at the School of Life Sciences at Arizona State University. 'Wherever you are in the universe, water is the key. Without water, it all falls apart.'

So it makes sense that scientists would think about the habitable zone in terms of having lots of water, but in that case where does a habitable desert planet such as Arrakis fit into it? It turns out that its lack of water might actually prove to be its saving grace.

One of the things that makes *Dune* such an incredible novel is that Arrakis feels like it could be real. The six years that Frank Herbert dedicated to working on the first *Dune* novel (Herbert wrote six novels set on Arrakis – *Dune Messiah* (1969) and *Children of Dune* (1976) directly continue the epic tale told in the first novel, while *God Emperor of Dune* (1981), *Heretics of Dune* (1984) and *Chapterhouse: Dune* (1985) are set thousands of years later) weren't just spent writing the story; he was fascinated by its ecology, and painstakingly worked out what kind of planet Arrakis was in the past, how it became the world it is in the first novel and how it would change and evolve in later novels. There are very few fictional planets that are as richly described, or as scientifically plausible, as Arrakis. As described by Baron Harkonnen in the novel, 'From sixty degrees north to seventy degrees south – these exquisite ripples. Their colouring: does it not remind you of sweet caramels? And nowhere do you see blue of lakes or rivers or seas. And these lovely polar caps – so small. Could anyone mistake the place? Arrakis! Truly unique.'

The level of detail in *Dune* has had a tremendously positive effect on generations of SF readers and writers, and scientists themselves. Paul McAuley describes *Dune* as being 'very influential when [he] was a teenager'. His contemporary Alastair Reynolds says that he was 'greatly impressed by *Dune*' as a young adult and that a recent re-read after going to see the 2021 film version showed that it still stood the test of time. It's certainly my favourite SF novel, but perhaps the most telling testimony is from Cambridge University's Claire Guimond, who we first met in Chapter One.

'I started getting into studying planetary science because of the first *Dune* book, which has a whole appendix on the imperial planetology science stuff, which I thought was so cool,' she says. 'I couldn't believe that it was so well thought out, and talking about how to completely change a planet from being in a desert state to a completely different state. I thought about that a lot and ended up getting into planetary science. I don't know if *Dune* was the only reason why. But I've always been really interested in Arrakis as a planet.'

In 2011 a quartet of planetary scientists – Yutaka Abe and Ayako Abe-Ouchi of the University of Tokyo, Norm Sleep of Stanford University and Kevin Zahnle of NASA's Ames Research Center – published a research paper called *Habitable Zone Limits for Dry Planets*. In this paper they figured out how habitable desert planets could be, and as a case study they chose a fictional world, Arrakis, which they describe as resembling 'a bigger, warmer Mars with a breathable atmosphere'. One imagines that they too were inspired by the strength of Herbert's magnum opus.

To understand why they think a planet like Arrakis can be habitable, we first need to understand why Earth is habitable, and that means getting to the bottom of this concept that we call the habitable zone. In our solar system, Earth sits slap-bang in the middle of the habitable zone, but more interesting perhaps is how far the habitable zone extends, both inwards towards the Sun and outwards towards Mars and Jupiter, where we can see general trends the closer one gets to the Sun, or the further away one is moved.

Earth's orbit around the Sun is slightly elliptical, meaning that it can get as close to the Sun as 147.1 million kilometres (which occurs every January) and as far as 152.1 million kilometres (every July). The average distance is 149.6 million kilometres, and astronomers use this to define the 'astronomical unit (AU)', a yardstick for measuring distances on a planetary-system scale. Earth is 1 AU from the Sun, Venus is 0.7 AU and Mercury 0.4 AU. Moving outwards, Mars is 1.5 AU, the Asteroid Belt is between 2 and 4 AU and Jupiter is 5.2 AU from the Sun.

Conservative estimates suggest that our solar system's habitable zone stretches from 0.95 AU to 1.01 AU. This is based on the fact that were Earth any closer, the extra warmth from the Sun would evaporate our oceans, dumping vast amounts of water vapour (a potent greenhouse gas) into the atmosphere, where it would kickstart a runaway greenhouse effect. This would rapidly raise temperatures so high that the remaining oceans would boil away, leaving Earth much like Venus is today: sweltering, dry and inhospitable. On the flip side, move Earth away from the Sun and all the liquid water turns to snow and ice, and our planet freezes over.

However, Earth has a built-in thermostat that can turn up the temperature when things start to chill and turn it down when things get too hot. We know from our struggles with human-caused climate change that carbon dioxide is another potent greenhouse gas, but it's belched out by not only cars and industry, but volcanoes too. Ordinarily, rain will draw the natural influx of carbon dioxide out of the atmosphere as carbonic acid, which can dissolve silicate rocks and interact with the carbon to form, for example, carbon silicate. This runs off into rivers and then into the ocean, where it is put to use by small marine creatures who incorporate the carbon silicate into their shells. When the sea organisms die, their remains sink to the seafloor, where, over time, subduction drags them into Earth's mantle. This subduction is instigated by plate tectonics – the movement of continental plates that results in earthquakes, volcanoes, the lifting of mountains and the Mid-Atlantic Ridge. In short, the carbon that was in the atmosphere is removed, at least until it is spewed back out again by volcanoes, and without the carbon dioxide in the atmosphere less heat is retained and the planet is able to cool. It's this natural process that is being circumvented by the burning of fossil fuels injecting additional greenhouse gases into our atmosphere.

When the temperature is colder, the level of precipitation is reduced, and carbon is no longer washed out of the atmosphere in any great quantity. Instead it remains, and volcanic eruptions steadily build the atmosphere's carbon dioxide abundance back up, warming the planet. So, if we were to move Earth further away from the Sun, this mechanism would kick in. Estimates suggest that Earth could remain habitable out to 1.65 AU – that's beyond the orbit of Mars, which is an intriguing idea that raises all kinds of questions about the Red Planet that we'll come to soon enough.

Yet desert worlds can circumvent some of these processes, and the quartet of scientists (Abe, Abe-Ouchi, Sleep and Zahnle) suggest that this is possible because of the lack of water on planets like Arrakis. Now hang on, didn't we say that the presence of water was key to habitability? It's true, but we didn't specify how much water. As we shall see in the next chapter, some exoplanets have so much water that they could be covered in a global ocean hundreds or even thousands of kilometres deep. Desert planets lie at the opposite extreme; they can still have some water, but crucially not enough to succumb to the same fate as Earth should our planet move a little closer to, or a little further away from, the Sun.

Not having much water means that there's not as much water vapour to fill the atmosphere if a planet is closer to its sun. This means that a runaway greenhouse effect can't really happen. And if you move a desert planet further from its sun, then there is not enough water available for the planet to freeze over and turn into a giant snowball, as would happen with an ocean world (which we'll learn more about in Chapter Five). Of course, there are limits

– there will come a point when a desert planet is too close to its sun, and thus too hot for life, or too far from its sun and too cold. The point is that the habitable zone for ocean worlds is actually quite narrow, whereas for desert planets it is much wider. A wider habitable zone means that more planets can fit inside it, and hence the more habitable desert planets there could be compared to the number of habitable water worlds.

## Common worlds

In 2009 NASA launched the wildly successful Kepler Space Telescope. Its mission, just like that of the *Enterprise*, was to discover strange new worlds: exoplanets of all sizes and distances from their parent star. Before Kepler launched, astronomers had discovered a few hundred exoplanets. Kepler found thousands.

With this cornucopia of extraterrestrial environments, astronomers were able to begin building some exoplanet statistics, and what those statistics told us was revelatory: on average, one-in-five Sun-like stars in our Milky Way has a planet or planets about the same size as Earth, orbiting in the habitable zone.

Although the distance of the habitable zone varies depending on how bright and hot a star is, Kepler's statistics are based on the traditional criteria for the habitable zone – you know, having lots of water. Yet the true number of potentially habitable worlds could be much larger because desert planets circumvent those criteria, and planets like Arrakis may greatly outnumber planets like Earth. If we're going to look for life in the universe, perhaps we should look for desert life.

Admittedly, these are all broad brush strokes. Not every planet in the habitable zone will be habitable – that's just the first step; the rest depends on the individual characteristics of a given planet. *Dune* was published three decades before the first exoplanets were discovered, so given what we know now, is Arrakis realistic enough to have the 'right stuff'?

If a desert planet is going to be habitable, it does need at least some water, somewhere. Arrakis is described as having aquifers and occasional morning dew, particularly near the polar regions, but an important facet of the story is that of the character Kynes, an ecologist who, along with his late ecologist father, realizes that, long ago, Arrakis was a wet world. The novel describes how he (Kynes is changed to a woman in the 2021 film adaptation) discovers a salt pan, left behind by water that once flowed openly on the surface, and he dreams of returning Arrakis to its previous environment by terraforming it. Yutaka Abe and company were correct in describing Arrakis as being like Mars, for the Red Planet too was once a wet world, billions of years ago. Today on Mars we see ancient shorelines, river deltas and floodplains filled with salts, sediments and aqueously altered minerals including goethite and gypsum, commonly found on Earth, and clay minerals such as phyllosilicates, which on Earth make up 90 per cent of our planet's water-drenched crust. One would expect a geological survey team on Arrakis to find many of these hydrated minerals.

Mars lost most of its water to space. It has the misfortune of being a small planet, only 6,794 kilometres wide, which is about half the size of Earth (our planet's diameter is 12,742 kilometres, or 4,220 mi.). This means that its gravitational field is weaker than Earth's, and Mars's interior also cooled much faster as the heat leaked away. As the molten core cooled, it stopped spinning, stopping Mars's magnetic dynamo effect dead in its tracks, and any global magnetic field Mars had dissipated. Combined, the low gravity and dying magnetic field were disastrous for the Red

Planet. With gravity only having a loose grip on the atmosphere, and no magnetic field to deflect the charged particles riding in from the Sun on the solar wind, both the atmosphere and its water vapour began to be stripped away. This process continues to this day, as has been observed by a NASA spacecraft in orbit around Mars, called MAVEN, the Mars Atmosphere and Volatile EvolutioN mission. What water didn't escape turned to ice and now resides at Mars's poles, or just beneath the surface as permafrost extending down to mid-latitudes. Today, Mars is a desert world – but a cold one, with temperatures averaging around −60 degrees Celsius.

Arrakis is not cold. The heat, especially around midday, is intense. The most habitable regions are the highlands near the poles, where the planet's capital, Arrakeen, is located. The tropics and equatorial regions are too hot for any human life. The nomadic human tribes of the Fremen, the native people of Arrakis, live in a multitude of underground and cave dwellings, or sietches, where they are sheltered from the sun. Outside Arrakeen, humans survive on Arrakis partly thanks to technology. For instance, a 'stillsuit', which captures all water leaving the wearer's body and recycles it, is vital.

With the atmosphere so dry it is described as having a moisture deficit, and any water that can be absorbed into it will readily do so. 'That's the reason for the stillsuits, to cover as much of the body as possible to stop moisture from being drawn out and into the atmosphere,' says Alexander Farnsworth, a climatologist at the University of Bristol. 'They don't do it in the films [the cinematic adaptations of *Dune* by director David Lynch in 1984 and Denis Villeneuve in 2021 and 2024], but the books talk about the stillsuit covering the whole face as well, because a lot of moisture comes out from breathing.'

Farnsworth, along with two other scientists, Bristol's Sebastian Steinig and Michael Farnsworth of the University of Sheffield, have modelled the climate of Arrakis to see whether it could hold up as a real planet.

Climatologists at the University of Bristol have a tradition of modelling fictional worlds. Middle Earth from *The Lord of the Rings* series and Planetos from George R. R. Martin's *A Song of Ice and Fire* novels (better known as *Game of Thrones* to television viewers) have been previous subjects.

'The beautiful thing about science fiction and fantasy is that it asks questions that probably wouldn't be asked scientifically by anyone else,' says Farnsworth. These questions include, just how do you get a planet with the exceptionally long summers interspersed by intense winters, à la *Game of Thrones*? Or what would the weather be like on a planet where every square centimetre of the surface is covered by a global city, like the planet of Coruscant in the *Star Wars* saga?

Arrakis offers unique challenges. 'Could you actually live there, and what would be the real problems that large human populations might experience – how would they adapt, how does the planet become a home-world for people who are intrinsically like us?' Farnsworth asks rhetorically. 'These broad, blue-sky questions are of really big interest to us.'

That's because although Arrakis is a fictional world, in our own future it may become startlingly relevant to the fate of life on Earth.

## Living on a desert planet

Humans are not the only life on Arrakis. There are the giant and feared sandworms, of course – the gigantic mega-predators that are often what people remember most about *Dune*. In the story, it is implied that the sandworms were cultivated by the Spacing Guild, since the larval stage of the sandworms produces the spice melange, a life-enhancing drug that is the most sought-after product in the entire universe owing to its ability to unlock a heightened awareness of the universe in those who take it, allowing the Spacing Guild to find safe paths through space–time for starships.

'Frank Herbert famously wanted to make a case for hydro-ecology being important, so he came up with this whole thing about how the sandworms were being exploited,' says McAuley. 'They were at the base of [how minerals were distributed], they shaped the planet in the way elephants shape the savannah.'

The sandworms are a conceit in the story of *Dune*. In reality, Dale DeNardo points out the absurdity of having large, predatory creatures on such worlds: 'I'm always amazed about how all these fictional planets, no matter how barren they are, have all these mega-predators that feed on lots of people,' he notes. 'Mega-predators need enormous water supplies [not sandworms, however – water is anathema to them], they need energy sources, and you wonder, what are they eating?' Well, according to *Dune*, the sandworms eat anything, including gigantic machines known as spice harvesters, which can't be great for the worms' digestion. Joking aside, if life is to survive on a desert planet, it must adapt itself to some pretty harsh conditions, but we see in Earth's own deserts how this is possible.

'Animals can survive in the desert by being economical, by using very little water, and by being able to store water reserves,' says DeNardo. 'They can go months without drinking, and they are able to tolerate dehydration.' DeNardo cites the example of the Gila monster, which sounds otherworldly but is actually a venomous lizard that lives in the Sonoran Desert in the southwestern United States and Mexico, including DeNardo's home state of Arizona, also home to the river basin after which the Gila monster is named. The Gila monster has been carefully studied by DeNardo as a unique example of how life can survive even the most inhospitable desert environments.

When water is scarce, as it is in a desert, many carnivores such as the Gila monster get their water from their prey, since about three-quarters of any animal is water. The chance to eat is a rarity in the desert, so it is vital that animals are able to store for long periods any water consumed. Gila monsters do this by 'using their urinary bladder as a canteen', says DeNardo. Gila monsters have the ability to store water there and reabsorb when necessary, leaving behind nitrous waste that solidifies. Snakes do a similar, but less effective, trick: 'They don't have a bladder, but they can store water in their colon,' DeNardo explains. 'Just imagine that: if they have to defecate, then there goes all their water.'

There is a downside to carrying water within yourself – it makes you heavy, so you'd better have a defence against other predators because you won't be running away from them very quickly. Gila monsters have a venomous bite, for example, and toads have poison in their glands.

Other desert animals create water metabolically. For example, the kangaroo rat – a small, hopping rodent that lives in the deserts of the United States – survives almost exclusively on its metabolic water, but to accomplish this it must have a voracious appetite (they live mostly on seeds) to get enough calories to burn to produce sufficient water as a by-product. In *Dune*, a very similar rodent – the 'kangaroo mouse' – exists in a near-identical ecological niche (we see one in the Denis Villeneuve films, and Paul Atreides chooses his Fremen name, Muad'Dib, after the kangaroo mouse). The idea for *Dune* came to Frank Herbert while he was researching a journalistic article about the Oregon dunes in the late 1950s, and he researched deserts quite carefully for the novel, leading to many of the ecological characteristics of Arrakis having origins in analogues found on Earth.

So, when it comes to water, evolutionary adaptations have upsides and downsides, and animals must fill their ecological niche with efficiency if they are going to survive. What's most interesting to DeNardo is that all these weird and wonderful tricks life can use to survive in the desert tend to be repeated by evolution across the world – and quite possibly across other worlds too, such as Arrakis.

'Regardless of the desert, whether it's the Sahara, or the Sonoran, or the Mojave, or the Gobi, we see these same types of adaptations,' says DeNardo: 'All the methods are there, just the names change, no matter what desert you're in, even potentially on other planets. There are different combinations that can make things work, but there aren't that many different combinations. Even exceptions are often duplicated, reflecting similar niches that animals live in that lead to the evolution of an atypical strategy. As you can tell, I'm pretty fascinated by this!'

Temperature is also a big issue for life to deal with on Arrakis, as it is in many deserts. It is implied that on Arrakis, summer temperatures can reach 70 degrees Celsius, and it's not that much cooler where Arrakeen is located, with a reported near-constant 27-degrees-Celsius temperature difference between the equator and the poles.

We have to fudge the details here a little bit. Arrakis orbits the star Canopus, which is a real star that observers in the Southern Hemisphere can see unaided in Earth's night sky, in the constellation of Carina. It's quite bright, the second brightest star in the entire night sky, second only to Sirius. Located

310 light years away from us, Canopus is an F-type star, which means that it is hotter and more massive than our G-type Sun. Indeed, it shines with an intrinsic luminosity 13,300 times greater than our Sun, and is eight times as massive, with a diameter of 98.8 million kilometres (61 million mi.). That's huge – our Sun is 'only' 1.39 million kilometres (865,370 mi.) across. This all means that Canopus's habitable zone is much further from it than Earth is from the Sun, and Arrakis would have to be correspondingly much further out from Canopus to be habitable, even further than Pluto is from our Sun, so far out that one year would last 430 Earth years. This doesn't quite mesh with how Arrakis is described in the *Dune* novel series, which is at a more Earth-like distance from its star.

Still, orbital period aside, the main point is that Arrakis is hot and dry, but we've seen how animals could adapt to the lack of moisture. While there are certainly physiological limits to what humans can put up with in terms of environment, technology can mitigate some of it. So how does the temperature and lack of water affect Arrakis's climate? That's what Alex Farnsworth wanted to find out.

Into his climate simulation went information on the surface topography of Arrakis (based on maps drawn by Frank Herbert, showing highlands in the mid-latitudes and polar regions and dune seas in the tropics and at the equator), the planet's orbital period and the shape of that orbit (Arrakis's orbit is near-circular; a highly elliptical orbit can lead to the long winters/summers seen in *Game of Thrones*), and the composition of Arrakis's atmosphere, which is described in the novel as being 75.4 per cent nitrogen, 23 per cent oxygen, 0.52 per cent argon, 0.5 per cent ozone, 0.23 per cent carbon dioxide and the remaining 0.15 per cent being trace gases. That's not too different to the make-up of

Earth's atmosphere (78 per cent nitrogen, 21 per cent oxygen, 0.9 per cent argon and 0.1 per cent trace gases), but one notable difference is the quantity of ozone on Arrakis. On Earth, ozone makes up just 0.00006 per cent of the atmosphere. In *Dune*, the excess ozone is produced by the sandworms.

Ozone is crucial because it protects against harmful ultraviolet light from the Sun, but ozone is also 'a massive greenhouse gas', says Farnsworth: 'Ozone's ability to warm the atmosphere is 65 times greater than carbon dioxide. If you put the same amount of ozone and carbon dioxide into the atmosphere, the ozone would warm the world 65 times more than the carbon dioxide. So that's why Arrakis is so hot.'

Indeed it is, and not necessarily in the places you would expect. On Earth, and on Arrakis in the novels, the hottest regions are the equator and the tropics, and the cooler zones are found in the mid-latitudes and the poles. That's the intuitive scenario that you would expect. Except Farnsworth's climate model found that on Arrakis, in real life, it would be the other way around.

The model showed that the warmest months in the tropics would reach 45 degrees Celsius, and during winter the temperature wouldn't drop below 15 degrees Celsius. That's in line with temperatures in the tropics on Earth. In the mid-latitudes, however, the temperature differential is far, far greater: 70 degrees Celsius on the sand in the summer, −40 degrees Celsius in the winter and the poles could drop as low as −75 degrees Celsius. Animal life as we know it, without technology and sheltering in purpose-built habitats, could not survive such a temperature range. If Arrakis was a real planet, Arrakeen would be built near the equator, not the north pole.

This reversal of temperatures is the result of cloud cover, or a lack thereof. Just like on Earth, Arrakis's equatorial regions receive the greatest amount of solar energy, but the sky is completely clear. Sand is highly reflective; scientists call this quality 'albedo', although there are different types of sand, made from different materials, with different grain sizes and different exposure to moisture (wet sand being darker than dry sand, as anyone who has been to the beach will know), and each with different albedos. In their climate model, Farnsworth's team adopted an albedo typical of sand in the Sahara, which is measured at 0.32, meaning that it reflects 32 per cent of all solar energy incident upon it. Reflecting it back out into space helps cool the reflecting surface, but of course, on Arrakis, as on Earth, the atmosphere can get in the way and trap some of the heat. This requires greenhouse gases; as we've seen, there's a lot of ozone on Arrakis. However, at the equator and the tropics, there's not much water vapour and very few clouds. The poles and mid-latitudes are a different story, with greater atmospheric water content and hence more clouds, and since water is also a potent greenhouse gas, the atmosphere at higher latitudes is able to trap more heat in the summer than winter.

With temperatures such as these, the polar ice caps described in the *Dune* novels would not survive in reality. They'd melt quickly in summer, with not enough snowfall in winter to replenish them.

## Stormy worlds

Another feature of Arrakis is the impressively violent 'Coriolis storms' – a bit of a nonsense name, says Farnsworth – which are effectively hurricanes combined with dust storms that can envelop large regions, or sometimes even the entire planet.

'Our model gives hints that such large storms could exist, but they would be confined mostly to the tropics and mid-latitudes,' says Farnsworth. 'The model actually goes quite a long way in suggesting that what is written about the storms in the novels is a possibility; there's a bit more nuance in how fast the winds are, how dangerous the storms are, how hot things are, but by and large, the model does agree with the world that Frank Herbert envisioned.'

Perhaps the closest analogy to the Coriolis storms in our solar system are the dust storms of Mars. The first spacecraft to orbit Mars, *Mariner 9*, arrived at the Red Planet in 1971. Rather than being greeted by a world of craters, mountains, deep valleys and vast dune fields, *Mariner 9* encountered a planet enveloped in its entirety by a global dust storm. It took months for the storm to clear.

Giant dust storms are a regular occurrence on Mars, particularly during Martian southern summers. This is because Mars's orbit is less circular than almost all the other planets (the orbits of all the solar system's planets are ellipses, but Mars is more so, second only to Mercury), and the point in its orbit when it is closer to the Sun, and therefore receiving more heat, is during summer in Mars's southern hemisphere (this also happens to be the case for Earth too).

However, we only have a general idea about what drives Mars's dust storms. 'One of the big questions that we have about Mars is, how do the big dust storms start?' asks Claire Newman, a planetary atmospheric scientist from Aeolis Research.

While the details are at best vague, what we do know is that crucial to the creation of dust storms is Mars's heat imbalance. When the Sun's warmth reaches Mars, it can either be absorbed or reflected back into space by the surface or the atmosphere – it all depends on the thermal properties of the materials in the surface and the atmosphere, and their albedo. On Earth the energy imbalance is very slight, with no more than 0.4 per cent difference in the amount of heat absorbed versus the amount re-radiated into space. (It's interesting to point out that Earth's energy imbalance would be a lot larger, and Earth a lot colder, without natural greenhouse gases in the atmosphere absorbing the Sun's heat, but now we have to contend with human-caused emissions of greenhouse gases destabilizing this balance.)

Mars is different. Between its northern and southern hemisphere during dust-storm season, the energy imbalance is 15.3 per cent, with more energy being absorbed, and this is what drives the dust storms. As the surface warms, the air immediately above it also warms, and, as we know, warm air rises. On Mars, rising warm air brings dust with it, while dust devils and even strong gusts of wind can loft more dust into the sky. All this dust is left hanging in the air to become the raw material for dust storms, but to have any kind of storm, on Earth, Mars or Arrakis, you first need winds.

Winds are generated by differences in the amount of surface heating at different elevations, and different albedos, and are then directed by the surface topography – around mountains, through valleys and so on. Mars has quite a bumpy surface compared to Earth, with more than 30 kilometres (18½ mi.) difference in height between its highest point (the towering extinct volcano Olympus Mons, which stretches to 21.9 kilometres (13½ mi.)) and its lowest point (the Hellas impact basin, a giant crater 8 kilometres (5 mi.) deep), so there's quite a degree of differential heating. Even so, Mars's thin atmosphere inhibits wind speeds – a dust storm might typically blow at about 100 kilometres (62 mi.) per hour, so not quite hurricane strength, unlike Arrakis's 700-kilometres-per-hour Coriolis storms. With all that dust in the air, the sky would darken – indeed, the loss of sunlight is life-threatening to robotic missions to Mars that rely on solar power – and there would be a lot more dust in the air, but you wouldn't really feel the full ferocity of the storm raging past you. The Martian atmosphere is so thin that a dust storm wouldn't threaten to blow over a spacecraft, for example, which was the potential scenario that led to Mark Watney being abandoned on the Red Planet in Andy Weir's novel *The Martian* (2011).

So that's how smaller, regional dust storms get started, but why some grow to be planet engulfing and others do not is a bit mysterious. Clearly, an important factor is how much dust might be at any given location when a storm begins, and that dust may take several years to be replenished after being blown away by a storm.

'We can understand the processes really well, but if you don't know how much dust is at a particular location, it's really hard to be able to estimate it accurately,' says Newman.

'Mars is a great way to try and test our theories about dust lifting and sand motion,' she continues. 'We have decades of wind-tunnel research and fieldwork out in Earth's deserts, and it's a good question, are they the same processes on Mars? For example, do electrostatic effects have a bigger impact because Mars is drier? On Earth water is an important factor in how much dust gets lifted.'

## Earth's fate

Earth will not have water for ever, however. As the Sun ages, it grows hotter, and the band of the habitable zone will begin to move outwards. In about a billion years, Earth will slip inside the habitable zone's inner edge and become uninhabitable.

The predictions are bad, at least for Earth as we know it. You see, stars are big balls of mostly hydrogen gas. In the nuclear-fusion reactions that run hot in the Sun's core, hydrogen is converted into helium via a process called the proton–proton chain, whereby protons, which form the nucleus of hydrogen atoms, fuse together through a series of reactions to assemble helium nuclei, releasing energy in the process. It's this nuclear-powered energy that creates the light and heat that we receive from the Sun; as I write this in a garden on a lovely summer's day, the photons hitting my arms have come from a natural nuclear reactor approximately 150 million kilometres away.

Hydrogen is the lightest element in the universe. A hydrogen nucleus is just a proton, as we've seen. Its by-product in the Sun's nuclear furnace, helium, is made from two protons and two neutrons, so helium is about four times heavier than hydrogen. As the Sun ages, the helium ashes of the Sun's nuclear reactions build up in the Sun's core, causing conditions there to become denser, and hotter, which drives more nuclear reactions, which puts out more energy, which creates even more helium, and so on and so forth. So as the Sun ages, it grows hotter, and brighter and wider (as the flood of photons pouring out puff the Sun up). Early in its life, the Sun was only 70 per cent as bright as it is today. In 1.5 billion years' time, the Sun will be 15 per cent brighter than it is now, and in 5.5 billion years' time, when the Sun is 10 billion years old, should there be anyone left alive on Earth, they would have to face a Sun that is twice as bright as it is now, with a diameter 40 per cent wider.

In 1.5 billion years' time, the temperature on Earth from the brightening Sun will rise to such a degree that the oceans will begin to evaporate, leading to an atmosphere filled with water vapour. The extra water in the atmosphere will heat Earth even more rapidly, evaporating more of the oceans in a vicious cycle – a runaway greenhouse effect.

We can look to Venus to see what's in store for Earth. Venus, being closer to the Sun than Earth is, has reached the point of no return before Earth has, and has experienced its own runaway greenhouse effect that a growing body of scientists believe happened only in the past 500 million to 1 billion years ago. Though that sounds like a long time, it's not so long in geological history, and if true, then Venus may well have had oceans, and maybe even life, a billion years ago.

Not anymore. Venus's water has all gone, boiled into the atmosphere and then lost to space. Its atmosphere today is a dense shroud of carbon dioxide, and its surface temperature reaches 460 degrees Celsius. When Earth's water literally boils away in the future, it's assumed that all life on Earth will end around this period, but can Earth avoid this fate?

It's possible, say the team of Abe, Abe-Ouchi, Sleep and Zahnle, ever the optimists. The trick, according to them, is dodging the runaway greenhouse stage, which frankly is fatal to life. Instead, Earth could become a desert world, like Arrakis or Tatooine, where life could survive by adapting to the conditions just as we see life adapting to the deserts on Earth today. Understanding how this could occur might also reveal how Arrakis-like planets form.

Since a runaway greenhouse would be caused by evaporated water, Earth needs to lose much of its water to space before the effect takes place. On the face of it, that seems paradoxical – we're wanting to avoid evaporated water causing a runaway greenhouse effect, but the early stages of giving up Earth's water to space will require putting all the moisture into the atmosphere. However, the team did some modelling of various scenarios and found that an atmosphere full of evaporated water could actually increase the albedo of Earth's stratosphere – the next layer of the atmosphere up above the layer where all our weather occurs, the troposphere. So a moist stratosphere could actually reflect some of the Sun's heat, keeping rising temperatures at bay. At this stage the water is in the upper atmosphere, where it is exposed to solar ultraviolet light that breaks the water molecules apart into their constituent hydrogen and oxygen atoms, and the solar wind plucks away the lighter hydrogen atoms.

'Once Earth reaches this stage, the stratosphere begins losing a lot of the moisture from the atmosphere to space,' says Farnsworth. 'The result is that you get a very dry but thinner atmosphere, as you might see on Mars.'

There's a caveat – the rate of water loss has to be finely balanced so that the atmosphere is able to thin coincident in time with surface temperatures rising, so that the surface doesn't become too hot. But, with enough luck, Earth would end up a desert planet. It may not be the relative paradise it is today, but given that the choice is between 'a liveable desert world versus a dead desert world', says Farnsworth, a liveable desert doesn't sound so bad.

That's what Arrakis, Tatooine and all the other desert worlds are – liveable but certainly not comfortable. We can turn to real-world examples to see how life might survive in unending sand dunes with little water, and we can look to science fiction to extrapolate on that and then listen to scientists who assess how much of the science fiction is fiction, and how much could be fact. In *Dune*, Arrakis is special because of the spice. Nobody would go there otherwise. If ever we need reminding how fortunate we are to live on a planet that is just right – not too hot, not too wet, not too cold, not too dry – we just need to look at the example of the desert planets, which could vastly outnumber the Earth-like planets in the galaxy. We're lucky to have Earth, and we should never take her for granted.

# 4

# OCEAN WORLDS

The Martians invaded because they were thirsty.

That's right. In H. G. Wells's all-time classic *The War of the Worlds* (1898), the Martians cross interplanetary space in their artificial cylinders to conquer Earth because their planet, Mars, is drying up. On Earth they begin bending our planet to their will, the alien Red Weed stretching forth across the land towards abundant sources of water. It's a theme that has been followed up many times in science fiction. For example, in the 1980s TV miniseries *V* and its sequels, aliens arrive over Earth's cities in huge flying saucers a mile wide (beating *Independence Day* to that startling visual by over a decade), at first claiming to be on a mission of peace before their real intentions are revealed: they want humans to eat and our water to drink. Reptilian creatures in human disguise, they start sucking Earth's oceans dry because, they say, a planet full of water is a rare thing indeed in the universe, and in the vein of all dastardly authoritarian extraterrestrial invaders, they're here to take ours.

I suppose this trope began with Percival Lowell, an early twentieth-century American businessman and what back in the day was referred to as a 'grand amateur' astronomer – a wealthy man with little to no scientific training who could afford to build a large observatory (which is still in use to this day in Arizona; Pluto was discovered from there in 1930). Lowell was obsessed with Mars, inspired by the observations of the Milanese astronomer Giovanni Schiaparelli, who in 1877 observed a network of lines criss-crossing the Red Planet. Back then telescopes were not as sharp or as powerful as they are today, and what Schiaparelli was seeing were optical illusions on the fringes of his perception. Nevertheless, he was not to know, and he called these lines *canali*, simply meaning natural channels. Alas, Lowell and others in the English-speaking world who were particularly impressionable took this to literally mean canals, as in artificially constructed waterways. Lowell built his whole observatory in an effort to observe these canals, and through the telescope eyepiece he convinced himself that he could. He drew maps of them and concocted a fiction that he believed in: Mars was drying out, and the canals had been built by the Martians to direct water from the poles to their cities in a last-ditch effort to avoid drought.

Suffice to say, there are no Martians and there are no canals or cities. Professional astronomers of the time dismissed Lowell's theories. While his eagerness to believe would have gone down well with Fox Mulder as an X-File, Lowell's delusion is a lesson from which we can all learn about being careful not to draw extreme conclusions from the flimsiest of evidence, and being mindful of ignoring experts just because the facts don't square with an imagined truth.

The irony is, from a certain point of view there was some truth in Lowell's fiction. When spacecraft finally arrived at Mars in the late 1960s and the 1970s, first in orbit and then missions that actually landed, they found a barren and dead desert world scarred with channels left by the planet's long-exhausted rivers. Mars had dried up, but this process had happened billions of years previously, much of its water having leaked into space because Mars's weaker gravity and lack of magnetic field left the water, and the atmosphere as a whole, susceptible to being stripped away by the solar wind.

Mars's predicament is not a sign that water is rare in the universe. There is actually still water on Mars, locked up in ice caps and permafrost – not quite the plentiful supplies as envisaged in Arnold Schwarzenegger's SF epic *Total Recall* (1990), but it's there. Beyond Mars, we find water almost everywhere in the solar system. It's in frigid, permanently shadowed craters on our Moon, it's found as vapour in the atmospheres of the giant planets Jupiter and Saturn, and their icy moons harbour vast, global underground oceans where there could still be life, as yet undetected. The so-called ice giants, Uranus and Neptune, contain large amounts of water in their bulk mass, while further out is the realm of the comets, each one a ball of dirty ice. Elsewhere, water vapour has been detected in the giant molecular clouds that form stars and planets; it's been spotted in the atmosphere of several hot Jupiter exoplanets, and it's even been identified in a quasar from which light has been travelling through space for so long and across such vast distances that we see it and its water (specifically, 140 trillion times the amount of Earth's water) as they existed 12 billion years ago.

Water isn't rare. It's a molecule made from the most common element and the third most common element in the universe – hydrogen and oxygen, respectively. There's tonnes of the stuff, and tonnes of the raw materials to make even more of it.

'Water is just all around,' says Claire Guimond, who we may remember is a planetary scientist at the University of Cambridge: 'It's a very common molecule. If you look in the outer solar system, the things there are like half made of water.' Robin Wordsworth of Harvard University agrees, but he takes things one step further. 'Personally I think water worlds are something we should expect in the universe,' he says.

Water might be plentiful, but it might not always be evenly distributed. In the previous chapter we focused on desert planets such as Arrakis from Frank Herbert's *Dune*. Although Arrakis's water loss was the result of a geoengineering project to cultivate a habitat to take advantage of the giant sandworms that produce the spice, which is the most precious substance in Herbert's fictional universe, there are plenty of ways that a planet can lose its water. Mostly it's through atmospheric loss, the water stripped away by the onslaught of stellar radiation, just like what happened on Mars.

On the flip side, there will also be planets that have lots of water, perhaps too much water. Earth has a good amount, with 71 per cent of our planet's surface covered by the wet stuff (including ice) that at its deepest, in the Pacific's Mariana Trench, extends more than 10 kilometres (6 mi.) down. The total volume of surface water on Earth is a mighty 1.39 billion cubic kilometres (around 333.5 million cu. mi.). Suck all the water off the face of the Earth and you could form a sphere 1,385 kilometres (860 mi.) across with it.

Yet planetary astronomers think that Earth only has a medium amount of water, relatively. If their observations of the density of some exoplanets is correct, then there are worlds out there that could be entirely covered in oceans hundreds, maybe even thousands, of kilometres deep.

If SF frequently features the trope that water is rare, then to balance things out it also has a lot of stories about ocean worlds. Stanisław Lem's 1961 novel *Solaris* (adapted as a film twice, a Soviet-produced version in 1972 and a Hollywood attempt in 2003 starring George Clooney) features a planet with a global ocean, but in this case the ocean is a single living organism that the human scientists in the story attempt to contact. A similar scenario arises in Stephen Baxter's novel *Galaxias* (2021), wherein a single powerful intelligence has evolved as an integrated component of a global ocean, only this time the ocean entity has more sinister motivations. In David Brin's award-winning novel *Startide Rising* (1983), the action takes place on the planet Kithrup, where the starship *Streaker* and its mixed crew of humans and 'uplifted' dolphins (that is, with artificially enhanced IQs) set down for repairs. Kithrup is covered mostly by oceans, with metal-rich mountains popping out of the water to form islands. Arthur C. Clarke also got in on the act with *The Songs of Distant Earth* (1986): in the novel, humans have settled on a few scattered islands on the ocean planet Thalassa, but by the year AD 3800, the Thalassans have lost contact with Earth, only to be visited by a spacecraft from home – an event that brings disruption to Thalassan society. And in Christopher Nolan's 2014 film *Interstellar*, Cooper (yay for a heroic scientist–astronaut character called Cooper) and his crew first visit a world called Miller's Planet that has a shallow global ocean but huge tidal waves. According to the film's science consultant and executive producer, Caltech professor and Nobel Prize-winner Kip Thorne, these deadly tsunamis are raised either by gravitational tidal forces enhanced by the rocking (called 'libration') of the planet as it orbits close to a supermassive black hole, or by the tectonic effects of those tidal forces.

SF author Alastair Reynolds describes ocean worlds as belonging to the category of 'panthalassic', named after Panthalassa, the immense ocean that surrounded the supercontinent Pangaea 250 million years ago on Earth. 'I think it's such a good word,' he says. 'I use it to describe an ocean planet in the third book of my *Poseidon's Children* series [*Poseidon's Wake* (2015)], where most of the action takes place, with alien stuff going on as well.'

So panthalassic worlds are plentiful in SF, and we think that they are plentiful in the universe too, perhaps even outnumbering the Earth-like worlds with their mix of land and water. Why are some planets water rich, and others dry or water poor? The water-rich worlds do, in theory, have some diversity. The truly water rich contain a large percentage of water in their bulk mass, while purely ocean worlds are thought of as rocky planets with a surface covered by a deep ocean. The source of all this variation is a question we're still working on answering.

The first hints that water worlds really could exist outside of SF came as astronomers were still reeling from the shock discoveries of hot Jupiters and super-Earths, both types of worlds that do not exist in our solar system and that upended conventional wisdom about planets. Once it became possible to make radial velocity measurements to get an exoplanet's mass, and transit observations to get the radius of the same planet, it was possible to calculate that exoplanet's bulk density by dividing its mass by its volume, which is derived from knowing its radius. Some exoplanets are evidently gaseous; others are clearly chunks of solid rock. Yet a large proportion have densities that are better explained if a significant fraction of their mass is water, whether it be in a gaseous, liquid or solid state. It's important to point out that water has not been directly detected on these planets, merely inferred through their density.

Early modelling suggested that water worlds, with a significant fraction of their bulk mass being made from water, could far outnumber planets that only have a relatively thin covering of water on their surface, like Earth does. This is supported in a 2020 study led by NASA's Lynnae Quick, which found that one-quarter of worlds from a sample of 53 smaller exoplanets, ranging in size from about the diameter of Mars up to twice that of Earth, showed signs of being water worlds. Many of these worlds score highly on the Earth Similarity Index (see Chapter Two).

## Dynamic chaos

Maybe we shouldn't be surprised. After all, we have our own water worlds in the solar system. Not planets per se, but the icy moons of the outer giant planets. Among Jupiter's many moons orbit three – Europa, Ganymede and Callisto – that are each thought to harbour a global ocean beneath their thick icy crust. Around Saturn, the moons Titan and Enceladus among others also fall into this category, as perhaps do some of Uranus's and Neptune's moons, and even that distant dwarf, Pluto. Yet the moons were constructed from leftover ices after their parent planet's formation. How water-rich planets around other stars form remains an open question.

'We don't know whether or not a specific planet is going to end up with a lot of water because a lot of the reasons come down to random events,' says Claire Guimond. Even the source of Earth's water remains uncertain, although we are getting closer to a definitive answer. It is highly probable that water was brought to our planet by myriad impacts when the solar system was still young, during a time of what Guimond describes as 'dynamic chaos', but was the water brought by impacts of asteroids or of comets?

The answer depends on the isotopes. An isotope is a version of an atomic element with the same number of protons but a different number of neutrons within the atom. Water molecules are made from two atoms of hydrogen and one atom of oxygen. Some water molecules contain regular hydrogen atoms, made of a single proton with an electron, but other water molecules can contain a version of hydrogen called deuterium that has one proton and one neutron plus the electron. When deuterium is found within a water molecule, we call this flavour 'heavy water' (it's not really heavy – the mass of a neutron is exceptionally tiny), and Earth's water has a specific ratio of heavy water to normal water (the D/H, or deuterium/hydrogen, ratio) of 0.000156:1. The source of Earth's water must have the same D/H ratio, but so far measurements of water on different asteroids and comets have provided mixed, inconclusive results. This has not been helped by the fact that Earth's own D/H ratio has evolved over time, some of the lighter hydrogen being preferentially lost to space at a faster rate than the heavier deuterium, shifting the goalposts slightly. The D/H ratio of the water belonging to a planetary or minor body such as a comet depends on the temperature of the location in the planet-forming disc of gas and dust around the young Sun where that particular object was made. The majority of comets have D/H ratios that are too high to be the source of Earth's water. A carbon-rich variety of asteroids tend to fair better, but nothing is nailed on for sure just yet.

So maybe some exoplanets received their water from impacts too, and those with more water maybe were on the end of a much heavier bombardment than Earth received. Alternatively, perhaps they were born already possessing all that water, before moving closer to their star.

Though you won't see a signpost for it, or a dotted line on a map, there is a boundary within our solar system. Inside that boundary, towards the Sun, during the time that the Sun and planets were under construction 4.5 billion years ago, the temperature was warm enough that water only existed as a liquid or vapour. Beyond that line, which we call the snow line, the temperature in the planet-forming disc of gas and dust was cold enough for water to freeze out as ice. The snow line existed somewhere between the locations of Mars and Jupiter today, and things that formed beyond it tend to be rich in water-ice. Interior to the snow line, water vapour could have been accreted into the planets as they formed, but the inner planets, including Earth, were born hot, their surfaces covered with oceans, not of water but magma. According to experiments performed by Paolo Sossi of the Swiss university ETH Zürich, the magma ocean would not have evaporated the water to form a steamy atmosphere; rather, the water would have dissolved into the magma and become permanently locked up there. Any water that Earth was born with was probably locked up in the same manner.

We would expect other planetary systems to have also had a snow line, with any icy worlds beyond it being rich in water. If those worlds were to somehow move closer to their sun, where the temperature is higher, they would begin to thaw. And if the past twenty or thirty years of study regarding planetary systems has taught us anything, it's that planets don't stay in one location. Early in their lives, they can migrate.

Migration is thought to have happened in our solar system. We know this because asteroids and comets have compositions that indicate they formed closer to the Sun, but they now orbit much further away, evidently scattered there by the gravity of a wandering Jupiter. This migration also provides an explanation as to why Mars is so small and ill-formed compared to Venus and Earth: the gravity of Jupiter pushed the young Mars into a gap in the protoplanetary disc carved out by the marauder's gravity, and in this gap Mars could no longer accrete the additional material it needed to grow larger. The disturbance caused by Jupiter's wanderings could even have led to water-rich asteroids being sent on a collision course with Earth.

Beyond our solar system the concept of migrating planets explains hot Jupiters and the warm 'mini-Neptunes' that astronomers are discovering in large numbers – in fact, they're the most frequent type of planet being discovered today. Imagine a world like Neptune – built beyond the snow line, rich in ices – that migrates towards its star. Temperatures rise and, beneath a thick hydrogen atmosphere, the ices turn to liquid, forming a deep and possibly habitable ocean, despite the atmosphere having a temperature of up to 200 degrees Celsius. Nikku Madhusudhan, a professor of astrophysics and exoplanetary science at the University of Cambridge who developed this concept of hot water-worlds, calls these hypothetical planets 'hycean' worlds, which is a portmanteau of hydrogen and ocean. If such worlds were to exist, having migrated closer to their star than Earth, they would make a mockery of our notions of a habitable zone.

Alternatively, perhaps that thick hydrogen atmosphere is stripped away by a sleet of stellar radiation as the planet edges ever closer to its star, leaving behind a rocky core possibly covered in water.

Astronomers think that they can see evidence of this, says Caltech's Jessie Christiansen. When astronomers plot the sizes of exoplanets on a graph, there's a weird gap between about 2.4 times and 1.8 times the radius of Earth. 'Nature doesn't seem to like planets that are exactly twice the size of Earth, which is odd,' she comments.

Two things happen to create this 'hot Neptune desert' ('hot' referring to the planet being closer to its star, and hence warmer than our Neptune). First, the heat from the core of a young planet is able to energize the atmosphere and allows the planet's atmosphere to be liberated into space. Second, is a phenomenon called 'photoevaporation', which astronomers have seen in action. 'Young stars have toddler tantrums,' Christiansen says, describing how a star's early life can be very unsettled, flaring with exceptionally strong gusts of radiation: 'As a planet migrates in, the star is trying to blow away that planet's atmosphere. There's a critical mass above which a planet can hold onto its atmosphere and below which it is going to lose a huge chunk of it.'

As a result, the hot Neptune desert represents the gap from planets that are just too small to retain their extensive atmosphere in the face of their gusty star. In doing so, they lose their hydrogen and helium envelopes, but in theory should manage to retain their water. Therefore, this desert has been interpreted as marking the transition of a mini-Neptune to perhaps becoming a super-Earth-sized water world.

Perhaps the best example of this kind of world found so far is TOI-1452b, which was discovered in 2022 by NASA's Transiting Exoplanet Survey Satellite (TESS) mission. TOI-1452b orbits in the habitable zone around two stars – a binary star system (see Chapter Seven) 100 light years away from Earth. It's a super-Earth 1.67 times larger than our planet, with a density that suggests that between 22 and 30 per cent of its mass is made from water, probably surrounding a rocky core. This proportion of water by mass is similar to what we find on icy moons such as Europa and Titan. Could TOI-1452b have migrated in-system after it formed beyond the snow line?

In Paul McAuley's 2016 novel *Into Everywhere*, which is the second book in his Jackaroo series, he features an ocean world that is a stripped mini-Neptune. 'That idea came directly from reading about exoplanets,' says McAuley. 'I thought that would be very interesting because mini-Neptunes are common around other stars, but our solar system is a bit odd because it doesn't have a mini-Neptune.'

Why some of the apparently most common planets in our galaxy – mini-Neptunes, super-Earths and hot Jupiters – are absent from our solar system remains a mystery, but it is this absence from our neighbourhood that makes them feel a little more exotic and otherworldly. In other words, it means they are great settings for fictional alien adventures.

## Habitable ocean worlds

However water worlds and ocean worlds form, it seems that at the least, the concept of ocean worlds in SF has its roots in reality. But the degree by which SF planets are covered by water can vary. On some worlds every inch is submerged; on others there are small continents, or tiny archipelagos where some of the story's action can take place – presumably because unless your main character is a fish, your heroes are going to have to spend at least some time on dry land. But land is also essential for the habitability of an ocean world because it is a vital source of nutrients. And as we shall discover, this poses a problem for a lot of water worlds in SF.

How much land might a water world have? This is where Claire Guimond's research comes in. She has been studying a property known as dynamic topography, 'which happens on any planet so long as it has a convecting mantle', she explains.

Topography is the shape of the land – the hills and valleys, mountains and basins that sculpt a planet's surface. On Earth the topography is mostly driven by plate tectonics, as continental drift pulls plates apart to create deeper basins or pushes them against each other until they fold into mountains. However, we don't know how common plate tectonics is on other planets – while there are hints that there once was plate tectonics on some of the solar system's other bodies, Earth is the only one with conventional plate tectonics now. (I say conventional because there is evidence that Jupiter's moon Europa has a form of plate tectonics in its icy shell, but it's not the same as on Earth, which as we shall see plays a specific role in our planet's geology and habitability.) Guimond avoided the question of how common plate tectonics is by modelling how a planet's topography might dynamically develop when it has a so-called 'stagnant lid', which is the absence of moving plates. A stagnant lid is a planetary surface too strong and immobile to be broken up into plates floating atop the deep solid rock of the mantle. The solid mantle (the mantle is not a molten ocean of liquid magma, a popular misconception that even I had until writing this book) flows via convection very slowly over hundreds of millions of years. The difference between a planet such as Earth that has plate tectonics and a stagnant-lid world is that, on Earth, the plates on the surface move with the flowing mantle, whereas a stagnant lid does not flow. So on the chance that plate tectonics are rare, or at least not ubiquitous, Guimond's results are totally independent of them.

As anyone who has flown in a hot air balloon can testify, hot air rises. It's a similar mechanism inside a planet's mantle – hotter pockets of mantle rock will be less dense and will therefore rise through the mantle, convecting upwards until they reach the top of the mantle and push through the surface, nudging the rock upwards so that on the topside the surface deforms into hills, creating topography. Guimond found that the hotter a planet's internal temperature, the less viscous the mantle was, resulting in some complex dynamics that lead to less deformation taking place on the surface.

More massive rocky planets – the breed of world that we call super-Earths – would be more susceptible to this because their interiors are expected to be hotter than Earth's mantle, and because their mantle plumes wouldn't be able to lift their stagnant lids. Also, the larger the planet, the greater the surface area. All of this affects how deep an ocean will be, but it's give and take: a flatter planet means that it takes less water, and therefore a lower sea level, to completely drown the surface, but the wider surface area means that there's more space for the water to spill out onto, lowering the sea level even further and giving the opportunity for any hillier highlands to stick their heads above water and create dry land.

A super-Earth would probably need something extra to push up the ground and provide topography – extra-ginormous and extra-long-lived convective plumes rising out of the mantle and driving powerful volcanoes might do it. There are issues with this, however; a super-Earth's stagnant lid would increase the pressure on the mantle, raising rock's melting point and therefore suppressing the production of volcanic magma, while any volcanoes would have to work against the planet's stronger gravity. It's an open question as to whether a super-Earth can even have volcanoes.

### *Blueheart*

Without topography – without land – our hopes for habitable ocean worlds take a hit because land is essential for the regulation of two processes essential to the survival of life as we know it.

One is the carbon–silicate cycle. Over long timescales – hundreds of thousands to millions of years – the carbon–silicate cycle acts as a planet's thermostat. Carbon dioxide belched out by volcanoes warms the atmosphere, since we know too well that carbon dioxide is a greenhouse gas. The warmer the climate, the more it rains, and gradually the rain removes much of the carbon dioxide from the atmosphere. This reduction of a potent greenhouse gas cools the planet, and as we'll discover in Chapter Five, this simple process can bring about ice ages.

As the carbon dioxide rains out, it reacts with silicate minerals in the rocks on the ground in a process that, crucially, is accelerated in warmer environments. The reaction of carbon dioxide with silicates produces calcium, bicarbonate and silica, which run off into rivers and then into the seas, where single-celled algae incorporate them into their shells of calcium carbonate. These algae are at the base of the food chain in the ocean and are consumed by and incorporated into all manner of other aquatic life. When these sea organisms die, their remains sink to the seafloor, where eventually plate tectonics sees subduction drag them into the mantle as tectonic plates move over one another. (It's worth pointing out that even without life, the weathering by-products may eventually settle onto the surface anyway, so the process could still take place on lifeless worlds.)

In the mantle, the carbon in life's remains is slowly recycled and, after a long time, belched back out into the atmosphere through volcanoes, warming the planet once more.

It's difficult to see how an ocean world, without land and possibly without volcanoes, could maintain the carbon–silicate cycle. Without that planetary thermostat a world will eventually either freeze or dry up.

The other process connected to rainfall and life is the liberation of nutrients. 'You need land to liberate the right nutrients, for example all the phosphorus that your body needs is coming from the rocks, the crust, so you need to erode that,' says Guimond.

That erosion mainly comes from flowing water and rainfall. The nutrient cycle on land is pretty straightforward – life needs water, and when that life dies its remains decompose into organic matter that releases the nutrients back into the ground. The origin of nutrients in the oceans is primarily the land, with rainwater carrying the nutrients into the streams, rivers and ultimately the oceans. Topography is required to raise the land above sea level so that it can be eroded and weathered. On ocean worlds with no land, this cycle can't happen, and even if there are some small areas of land, the sheer amount of water will work against the process.

'An ocean world won't be that productive because it will have so much water that any nutrients will be diluted,' says Abel Méndez, who is a planetary scientist and astrobiologist at the Planetary Habitability Laboratory at the University of Puerto Rico at Arecibo, whom we first met in Chapter Two as the co-originator of the Earth Similarity Index. 'Any nutrients present will eventually sink to the bottom. The water would be purer than here on Earth, and the nutrients would be such a low amount they could only sustain microbial life, not animal life.'

This was the dilemma that faced the Scottish–Canadian author Alison Sinclair when she wrote her ocean-bound SF novel *Blueheart* (1996). It's a fascinating and enjoyable story of human settlers on an ocean world, and the conflict between those who want to change the planet to suit them and those who are changing themselves genetically to suit the planet, which is nicknamed the eponymous Blueheart. It is the fifth planet orbiting the star gamma Serpentis (Sinclair calls it 'Gamma Serpens'), which is a real Sun-like star about 37 light years from Earth (at the time Sinclair wrote the novel, the star was thought to be 42 light years away, but this distance was revised to 36.7 light years in 2007). It's a slightly warmer star than our Sun, and Blueheart enjoys a tropical climate, with 97 per cent of its surface covered by ocean.

Allow me a brief diversion to talk about exoplanetary naming conventions. Officially, all exoplanets are catalogued with their star's Latinized name followed by a letter. So, Proxima Centauri, in the constellation Centaurus, has three known planets orbiting it: Proxima Centauri b, c and d. Proxima 'a' is the star itself. Some exoplanets have been given proper names; for example, 51 Pegasi b, which if we recall from Chapter One was the first planet to have been discovered around a Sun-like star, has also been alternatively named Dimidium and Bellerophon, but frankly, nobody uses those names. Perhaps once we have telescopes powerful enough to provide direct images of exoplanets, showing land and sea and ice and vegetation, we'll stop thinking of these worlds as catalogue numbers and start using proper names. But back in 1996, when *Blueheart* was published, barely any exoplanets had been discovered, so this naming convention was still young. Sinclair calls Blueheart Gamma Serpens V, which uses the old SF trope of giving planets numbers (see also *Babylon 5*'s Epsilon III, *Star Trek: The Wrath of Khan*'s Ceti Alpha V, *Forbidden Planet*'s Altair IV and many others). More properly, Gamma Serpens V, which is in the constellation of Serpens, the Serpent, should be named gamma Serpentis f.

Anyway, back to *Blueheart*: early in the writing of her novel, Sinclair, who is a biochemist but who also studied and worked in medicine and neuroscience, realized that her ocean world had a problem. In a blog post, Sinclair talks about how Blueheart's ocean, being warmer and less salty than Earth's oceans on average, is therefore less dense and floats atop a deeper layer of denser water. The aquatic life on Blueheart lives in that top layer, but when that life dies its remains, along with the nutrients those remains contain, would sink right to the bottom of the dense layer. She raises an additional point that on Earth, deep water is mixed with surface water by winds that drive surface water away from coasts, allowing deep water to well up, but with no continents on Blueheart there are no coasts, and with barely any land there's no source of nutrients to replenish those that have sunk to the bottom. What to do?

Sinclair credits the SF and fantasy author Tad Williams with coming up with the solution – a false bottom in the form of a layer of floating forests with a thick mat of tangled roots that helps trap the nutrients.

You might then ask, how did the floating forest form if all the nutrients were on the ocean floor? This is not answered in the novel (in fact, most of this world-building is barely mentioned in Sinclair's story, forming only a backdrop that helped make the planet feel more real to Sinclair as she wrote a rounded portrayal of an ocean planet and its mechanisms), but the fact that the nutrient issue was even recognized as a problem scores Sinclair points.

Ocean worlds like Blueheart are not necessarily an extreme, because they could be quite numerous, but on a scale of water-based habitability with Earth in the middle, they are to the extreme right of Earth. On the flip side, to the extreme left with low water content are desert worlds like Arrakis; because they have no rain, and therefore no weathering can take place, no nutrients can be made available in the non-existent run-off. Two completely opposite planet types, but if either land or ocean is missing the result is the same.

'The sweet spot is to have land and ocean,' says Méndez. Something like *Avatar*'s Pandora is certainly a type that comes to mind when we think of that sweet spot fertile with potential. However, as we've alluded to already in this chapter and in Chapter Two, planets in that sweet spot might be rare. 'In the models I've been working on, there is only a very narrow window where you can have these intermediate land/ocean fractions like Earth,' says Guimond. Even if the probability of being an ocean world, a desert world or an Earth-like world was equal, it would still mean that two-thirds of habitable-zone Earth-mass planets would not be in that sweet spot.

## Can we find ocean worlds?

Identifying planets in the sweet spot is not going to be easy. Astronomers' current telescopic technology just doesn't have the ability. Consider that we obtain an exoplanet's mass and radius without even directly seeing the planet in most cases. If you took all 1.39 billion cubic kilometres of water off Earth, it would barely affect our planet's overall density. Alien astronomers using similar means as we have to study Earth would not be able to tell that Earth has oceans. Vice versa, it will be all but impossible to detect Earth-like oceans on an exoplanet simply by measuring the planet's density. Even in the cases where the measured density suggests a large fraction of a world is water, there's no way to tell whether it is surface water, how deep that water is and whether there's any land.

The next step in our exoplanet studies is analysing planetary atmospheres. Even though in most cases we can't see the planet directly, astronomers are able to use a technique called transit spectroscopy. When a planet transits its star, light from the star passes through the planet's atmosphere and is absorbed at specific wavelengths by different molecules. During the first twenty years of the twenty-first century, astronomers had used this technique with the Hubble Space Telescope and the infrared-seeing Spitzer Space Telescope to probe the atmospheres of several hot Jupiters, finding a few molecules such as helium and even water vapour. Now, with the 2021 launch of the far more powerful JWST, it has become possible to probe the atmospheres of these worlds with greater clarity, or to detect atmospheres of smaller rocky worlds and mini-Neptunes, which was impossible before.

The problem with transit spectroscopy is that it can only tell you about the content of a planet's upper atmosphere. The JWST couldn't directly detect a planetary ocean in this fashion, says JWST exoplanet scientist Knicole Colón. However, she suggests how the JWST could still provide very strong evidence of an ocean. 'If you see a saturated water atmosphere then that would presumably equate to a water world because it is so saturated compared to anything else,' says Colón.

If the sky on an exoplanet is clear of cloud, then 'there's also the idea that we could detect the glint of oceans,' says Guimond. This glint wouldn't be the kind of ocean glint you can catch on camera – as we've discussed, we currently lack the technology to take detailed photographs of such worlds. Instead, the glint would appear as an anomalous bright spot detectable as a bump in the planet's light curve, which is a graph that shows how its brightness changes over time as it orbits around its star.

Colón is sceptical that even a light-curve observation could be made with current technology; 'I think the ocean glint idea might be more viable if you have a high enough signal-to-noise,' she says. She's waiting for the next generation of space telescope, with at least an 8-metre-wide (26 ft) mirror – compared to JWST's 6.5-metre (20 ft) mirror and Hubble's 2.4-metre (8 ft) mirror – to be able to conduct direct imaging of exoplanets.

'We need a future mission to take those direct images to see whether there are oceans,' says Colón. Although even an 8-metre space telescope couldn't resolve an exoplanet as anything but a point of light, Colón points out that 'With direct imaging, you could ultimately do some clever things like seeing changes in surface emissions or albedo, and measuring their period allows you to get the planet's rotation rate. You could even look at continental distribution versus latitude and things like that. It's going to be a whole step beyond what it's possible to do right now.'

Of course, like the rest of us, astronomers can be impatient. The next generation of space telescope isn't going to launch until at least the 2040s and possibly later. That's a long time to be twiddling thumbs, so some astronomers have come up with innovative ways of circumventing technological limits. For example, NASA's Lynnae Quick suggests that some water worlds might be like big versions of Jupiter's moon Europa, or Saturn's moon Enceladus, both of which sport frequent geysers of water that squirt out into space as gravitational tides from their parent planets flex their interiors. Quick thinks that JWST spectroscopy would be able to detect such geysers.

Then 'there's a really great paper by Robin Wordsworth,' says Guimond. 'It's about detecting a sulphur species in an atmosphere that shouldn't be there if there's an ocean.' Wordsworth himself continues explaining the idea: 'We're interested in how you could use the upper atmosphere signal indirectly to say things about the presence of water on the surface,' he says. 'It turns out that whether you get sulphur dioxide as a gas in the atmosphere of a planet or not depends a lot on if liquid water is present, because sulphur likes to react with water and it all gets sucked down back into the oceans pretty quickly. So detecting sulphur dioxide would be a kind of anti-liquid water signature.'

Over the coming decade, the idea should be put to the test as JWST makes increasingly detailed studies of exoplanetary atmospheres. Wordsworth doesn't necessarily think he's going to find lots of planets with sulphur dioxide missing in their atmosphere, but the number of worlds that do have sulphur dioxide 'is going to be able to constrain the number of planets with oceans'.

Being able to constrain that number is incredibly important because it will put to the test the notion that planets drenched in liquid water are common. Should we find lots of rocky planets and mini-Neptunes with copious amounts of sulphur dioxide in their atmosphere, then perhaps liquid water oceans really are rare. If that is the case, then maybe the scenarios of *The War of the Worlds*, *V* and their contemporaries are not so far-fetched after all.

# 5

# WORLDS OF ALWAYS WINTER

Imagine a world where there is no respite from the cold, where winter creeps from the poles and heads mercilessly towards the equator, stopping just short of fully encapsulating the globe. Dividing the two white walls of ice and snow in each hemisphere is an equatorial strip of land and sea sandwiched between the encroaching frost, a temperate zone where humanity can exist. Even during 'summer' in this habitable strip, it never grows particularly warm; temperatures climb no higher than 5 or 10 degrees Celsius, and it rains a lot. Winter is drier, but very cold. The inhabitants have adapted to the conditions over millennia, but visitors to the planet find the climate intolerable.

The planet is called Gethen; it's from Ursula K. Le Guin's award-winning 1969 SF novel *The Left Hand of Darkness*, and it's one of the best-realized worlds in the science-fiction lexicon. In the novel, a human envoy called Genly Ai, hailing from a galactic civilization known as 'the Ekumen', arrives on Gethen to contact a long-lost human society that exists there. The humans of Gethen have evolved to be sexless, except for periods of high fertility called 'kemmer', where they can transition into a male or a female. Upon his arrival, Genly – a male from Earth, which is part of the Ekumen – is greeted by suspicion, paranoia, even disgust at his permanent gender. During the story Genly must set out with his sole ally on Gethen in a desperate trek across the ice.

Le Guin's themes of gender, nationalism, first contact and political intrigue are all as pertinent now as they were in 1969, if not more so. Perhaps that's a sign of good science fiction – stories that grow more relevant, not less, over time. From the point of view of our exploration of how fictional worlds match up to real-life planets, Gethen is extremely relevant.

In her novel, Le Guin suggests that Gethen's chilly climate is the result of low levels of solar irradiation and that perhaps Gethen is near the outer edge of the habitable zone. As the narrative describes, a further decrease in heat from Gethen's star of about 8 per cent would bring the two walls of ice together at the equator and slam them shut.

You can tell that a fictional exoplanet is a good one when you are reading the scientific literature and are able to recognize the science-fiction world in the scientific facts. In 2019 Adiv Paradise, an astrophysicist, climate scientist and self-confessed science-fiction fan at the University of Toronto, Canada, co-wrote a paper called 'Habitable Snowballs: Temperate Land Conditions, Liquid Water and Implications for $CO_2$ Weathering' with his University of Toronto colleagues Kristen Menou, Diana Valencia and Christopher Lee. Their paper described how their simulations showed that some exoplanets so cold as to be covered in ice could still support large, unfrozen areas of habitable land, perhaps reaching summer temperatures of 10 degrees Celsius. When I read that, I snapped my fingers in recognition – what they were describing sounded just like Gethen!

When I spoke to Paradise, he revealed that he'd not read *The Left Hand of Darkness*, but I couldn't help but be struck by the similarity of Le Guin's Gethen and the simulated worlds realized by Paradise and his cohorts. 'It does sound pretty similar to what Ursula K. Le Guin has in her novel, where there is this belt around the equator that's not completely frozen,' says Paradise.

Let's rewind a little bit. Paradise studies so-called snowball planets – worlds like Hoth in *The Empire Strikes Back* (1980). The visuals stick in the mind: a planet completely covered in white, with hulking AT-ATs stomping over the ice and swatting away rebel snow-speeders. Seen from orbit on board an Imperial Star Destroyer, every square metre of Hoth's surface appears to be frozen. A snowball indeed.

Such worlds are not just figments of our imagination. Although astronomers' ability to characterize the climates of exoplanets to that degree is currently lacking, we do know of one planet in the habitable zone that has completely frozen over: Earth. Yes, our planet has undergone several snowball events itself, including one between 2.4 and 2.1 billion years ago, and then two more 730 and 610 million years ago, respectively, encapsulating an era referred to as the Cryogenian period. Even recent evidence from Mars of the remains of glaciers close to the Red Planet's equator suggests it too may have been covered in ice at some point in the past 2 or 3 billion years.

So Earth and Mars could have been very much like Hoth or Gethen at several points in their history. The intention of this book is to take fictional planets from SF and ask whether there could be real planets like them, but in this case there's more to be gained from flipping the question on its head. Which of those two fictional worlds, Hoth or Gethen, was Earth most like during its Cryogenian period? It is an important question because, evidently, life somehow survived those snowball stages, and we'd quite like to know how it did it – and what caused the planet to ice over in the first place. Understanding this could help tell us about the potential habitability of other worlds in the cosmos, too.

## Total snowball

We're familiar with the ice ages, the most recent being between 115,000 and 11,700 years ago, when modern humans emerged from this cold spell. Ice ages, however, are not snowball periods. An ice age sees the rapid advance of glaciers down to the mid-latitudes, with global temperatures dropping by about 5 degrees Celsius on average, but much of the planet still remains temperate. Furthermore, ice ages are fairly regular occurrences. On geological timescales, Earth moves through different glacial and interglacial periods, which are the result of climate forcing by a trio of effects that cumulatively produce the Milankovitch cycles. Named after an early twentieth-century Serbian scientist, Milutin Milanković, the Milankovitch cycles have a significant impact upon the climate of our planet.

First, the interplay of interweaving gravitational forces from the Sun, the Moon and even the other planets in our solar system can act to change the eccentricity of Earth's orbit around the Sun. As we've seen in previous chapters, Earth's orbit is slightly elliptical, rather than circular, but the effect of the gravity from the solar system's other bodies can reshape how elongated this elliptical orbit is, with two distinct cycles with periods of 100,000 and 400,000 years, respectively. This means that, at times, Earth's 'aphelion' – the most distant point from the Sun in our orbit – is a little further away on average, or a little closer on average, to the Sun than at other times, and clearly this affects the amount of heat our planet receives.

The second factor is the angle at which Earth spins. Our planet is tipped over by 23.4 degrees relative to the plane of the orbits of the planets, which we call the ecliptic. This angle can vary, however, between 21.5 and 24.5 degrees over periods of

41,000 years, again pulled by the gravity of other celestial bodies. Earth's tilt shapes its seasons, since different hemispheres are slanted either towards or away from the Sun at different points in our orbit.

The third factor is the precession of Earth's spin. Like a child's spinning top, our planet wobbles on its axis, and precession is what we call the rocking of the axis of rotation. If you were to draw a line through the centre of Earth and out through the geographic poles, along the axis of rotation, then over a period of 26,000 years this line would appear to trace out a circle. This is why the Pole Star changes over time – currently that line, drawn northwards through the axis of rotation, points to the star Polaris, in the constellation Ursa Minor, the Little Bear. However, because of precession, the polar axis will move away from Polaris over time and point towards other stars. For example, in about 12,000 years' time, our north pole will point towards the bright star Vega in the constellation of Lyra, the Lyre, instead. Yet the consequences of precession are more far-reaching than simply changing the identity of the Pole Star; it also alters where in Earth's orbit the seasons in each hemisphere take place. Currently, northern summer takes place when Earth is near aphelion, which is when we're furthest from the Sun, and southern summer occurs near perihelion, when Earth is closer to the Sun in our elliptical orbit. Precession means that eventually this will change, and the amount of insolation each hemisphere receives will be affected. This effect is enhanced when coupled with the fact that the eccentricity of Earth's orbit can also change, along with our tilt.

Combined, these three factors can force our climate to change over hundreds of thousands of years. It's all entirely predictable, and there's nothing mysterious about it. Even during a typical ice age, like the most recent one that ended about 11,700 years ago, there are still parts of the planet that are balmy.

Snowball events are different. Evidence for Snowball Earth – a term coined by the Caltech geologist Joseph Kirschvink in 1990 – first came to light when the remains of glaciers were found near the tropics, at far lower latitudes than typical ice age periods can account for. Some scientists suggested that perhaps Earth had been tilted over significantly more than its usual range of 21.5–24.5 degrees, which would have meant that the poles would have seen more of the Sun than the tropics and equator, but nobody could explain how Earth would have then righted itself. Instead, geologists and climatologists came to the conclusion that on numerous occasions, particularly those aforementioned periods 730 million and 610 million years ago, Earth became completely covered in snow and ice, from the poles all the way down to the equator, where the ice met and slammed shut.

'The snowball planets are interesting because that concept comes out of our own geology, where the geologists and palaeontologists have found this Snowball Earth period in our own history,' says NASA planetary scientist Alex Howe of the Goddard Space Flight Center.

Exactly what instigates a snowball event remains uncertain. It ain't the White Walkers from *Game of Thrones*, that's for sure. We of course expect planets far from their Sun, beyond the typical habitable zone, to be icy, but we wouldn't ordinarily expect it of worlds in the habitable zone that are close enough to the Sun to normally be temperate. In 2020 Daniel Rothman and Constantin Arnscheidt of the Massachusetts Institute of Technology put forward the idea that it's all down to a sudden decrease in sunlight reaching the planetary surface, mimicking the reduced sunlight that a planet too far from its star would experience. Rothman and

Arnscheidt estimate that over the course of 10,000 years, Earth would have needed a drop in sunlight of just 2 per cent to trigger a snowball event. Maybe supervolcanoes gradually fill the sky with ash, or dust is raised from an asteroid strike or even airborne algae that increase cloud formation, the extra cloud cover being able to reflect more sunlight back into space.

Stars can be variable in their brightness too. Although our Sun appears stable, other stars are not, and orbiting exoplanets might find themselves cycling in and out of snowball phases in time with their star's mood swings.

Another theory suggests that life itself – or more specifically, the evolution of photosynthesis – could be the trigger for a snowball event by sequestrating much of the carbon dioxide in the atmosphere. This in particular seems applicable with the deep freeze that is thought to have been inflicted upon Earth after the oxygenation of the atmosphere 2.4 billion years ago. Prior to that, there had been little free oxygen in Earth's atmosphere, but the evolution of microorganisms called cyanobacteria, which developed the pigment we know as chlorophyll, changed the world. Via the magic of photosynthesis, cyanobacteria in plants can take carbon dioxide from the atmosphere, mix it with energy from sunlight and metabolically convert it into biochemical energy with oxygen as a waste product. Carbon dioxide, as we know, is a potent greenhouse gas.

'There's speculation that plants evolving – or rather, photosynthesis evolving – may have triggered one of our snowball events by removing a lot of the carbon dioxide from the atmosphere,' says Paradise.

This reduction in planetary warming was akin to being stood around a bonfire, cosy and warm, and then taking your clothes off so that you get cold again.

Whatever the ultimate cause, at some point a tipping point is reached. Ice is reflective – it has a high albedo, which, if we recall, relates to how much sunlight a surface reflects. Ice reflects about 65 per cent of sunlight incident upon it back into space, and fresh snow as much as 80 per cent. The colder a planet grows, either from reduced sunlight or reduced greenhouse gases, the more ice develops, the more sunlight is reflected away, the colder the planet becomes, the more ice develops and so on – a vicious cycle of decreasing temperature. Once the ice reaches the tropics, where the Sun is high in the sky, it is able to reflect a lot more sunlight back into space than at the poles, where the angle of the Sun is shallow. This really accelerates the process, and once that critical point is reached, says Paradise, the onslaught of the ice is swift. 'It all happens very quickly once you hit the tipping point,' he says. 'The ice reaches the equator in a matter of years, and slams shut.'

The fact that there is still some life on Hoth – the creature that attacks Luke Skywalker at the beginning of *The Empire Strikes Back*, a kind of abominable snowman named a wampa, is native to the planet – suggests that it is in the habitable zone, and that not too long before the events of *The Empire Strikes Back*, it was able to support a complex biosphere that could sustain large mammals. Therefore, Hoth may have entered its snowball state only a few years or decades before Luke and the rebels encounter it. The wampas might be clinging on to the last vestiges of the food chain: if they originally lived in Arctic conditions at Hoth's poles, they might have been able to adapt to the cold conditions, migrating across the planet as the ice spread.

## Surviving the ice

Earth entered its snowball states before complex eukaryotic life had a chance to evolve greatly; it was still very much a world of algae and microbes. Perhaps such simple life found it easier to survive than large animals like wampas would have, but there were still significant challenges facing their continued existence, challenges that life evidently overcame, otherwise you wouldn't be here to read this book – nor I to write it, for that matter.

'A snowball state has happened two or three times in Earth's history, and it very clearly did not wipe out all life, and there actually isn't any evidence that it caused a decrease in biodiversity at all,' says Paradise. In fact, there's some suggestion that it prompted an increase in biodiversity and helped fast-track evolution as natural selection worked to give life that could adapt to the new conditions an advantage. Is it a coincidence that the Cambrian explosion, which saw a multitude of weird, wonderful and complex new lifeforms evolve on Earth, occurred just after the last snowball event ended 530 million years ago?

There are several hypotheses concerning how primitive life may have survived the ice. Scientists led by Maxwell Lechte of McGill University in Canada have studied ancient red rocks that were submerged in dark oceans with a ceiling of ice thousands of metres thick over their heads during the two snowball eras. The rocks' rich hue indicates that they contained a wealth of iron that rusted. For the iron to rust, the rocks must have been exposed to oxygen, which they should not have received much of while trapped under all that ice, unless there was a secret supply of oxygen to the oceans. Experiments performed on the rocks by Lechte's team found that those rocks that formed out in the deep sea rusted much less than the rocks that formed closer to the coastlines, where the ice entered the ocean. Lechte and his team compared the situation to the three-hundred-plus ice sheets in present-day Antarctica, which contain large amounts of trapped air that is passed into the glacial meltwater, which subsequently bubbles with oxygen and feeds the ocean around the ice. Lechte's team propose that similar streams of oxygen-rich meltwater could have helped oxygenize the submerged ocean near the coastlines. Besides rusting iron-rich rocks, this oxygen could also have kept microbial life alive.

There's another possibility, which brings us back to Gethen and its strip of temperate land and sea separating two white hemispheres, or perhaps Darkover, from Marion Zimmer Bradley's novel series of the same name, which also features a narrow, frigid but habitable equatorial strip between a wall of ice-covered mountains to the north and the sea to the south. Lest we forget the sixth planet of the sigma Draconis system in the first episode of the third season of *Star Trek*, 'Spock's Brain' (aired in 1968), which is also in the midst of an ice age with a barely temperate equatorial belt (the episode being great fun, but completely daft, with Spock's brain being stolen and Dr McCoy operating Spock's body via remote control until his brain can be put back in). What if during the two snowball events, Earth – like Gethen, Darkover or even the sixth planet around sigma Draconis – was not completely covered in ice but left with some naked land on which life could survive?

Adiv Paradise says that what determines how much of a planet becomes covered in ice is a combination of distance from the planet's star, how much carbon dioxide is present in the atmosphere, and atmospheric circulation, which turns out to be quite important in how and where ice sheets and glaciers form and expand.

'The reality is that glaciers do not appear out of thin air,' he notes. 'They are grown over time by snowfall that doesn't melt, and snowfall comes from moisture in the air, and on Earth most of the water vapour in our air comes from the oceans.'

This poses a potential problem for any planet that's looking to go 'total snowball': once ice starts to cover the oceans, it cuts off the supply of moisture to the atmosphere that is needed to enable it to produce snow and create more ice, and this is what leads to unfrozen regions.

Gethen's unfrozen region is at its equator, but in reality it need not be. Where land remains unfrozen depends upon where the oceans are on the planet and the global atmospheric circulation.

'What was surprising in our research was that we ended up with warm areas in continental interiors away from the equator,' says Paradise. For example, in their simulated snowball Earth, Paradise's team found a warmer region keeping large parts of North America unfrozen. Though not a snowball era, during the last ice age (115,000 to 11,700 years ago) northern Russia was mostly not covered in ice, despite being in a location where you would expect ice to be dominant, because the air currents tended not to carry a lot of water vapour over that area so there wasn't enough snow to create glaciers.

However, we might imagine a world that has most of its water at its poles and most of its land near its equator. Once the oceans freeze, the supply of moisture to the atmosphere is cut off, and then there's not enough snow to create glaciers over the equator, leaving a strip unfrozen. In the case of Gethen, Le Guin's maps describe one hemisphere as being mostly land with some seas, and the other hemisphere as being mostly chilly ocean strewn with floating icebergs and glaciers. Of course, ice covers a lot of both hemispheres, and it would require climate modelling to figure out whether atmospheric circulation could keep the equatorial regions free of ice.

As for Hoth, we don't really know what kind of planet that was before the ice came. If it was mostly land, then it would have been much harder to build vast ice sheets because glaciers are made from compacted snow, and as we have seen, snow is produced by precipitation of water from the atmosphere, and the primary source of the water content in the atmosphere is the oceans. A planet with more land than sea is going to struggle to produce enough snow to cover the planet with glaciers.

'We don't see much of Hoth in the film,' Paradise speculates: 'We only see where the rebels are, and there are mountains, but maybe those mountains originally formed archipelagos, and it is easier to build glaciers on them . . . The other option is that we don't know for sure how close Hoth is to its star. We know it's very cold where the rebels are. At night there might be dangerous temperatures, and we see that in the film, but as long as daytime temperatures are not too brutal, maybe –30 degrees Celsius – I live in Canada and that's an annual experience during our winters – then the rebels could survive there.'

Shutting off the supply of water vapour isn't all positive. Water vapour, like carbon dioxide, is a greenhouse gas – its presence in our atmosphere contributes to keeping our planet warm, so without it the air would be both drier and colder. The same goes for planets where the ice has slammed shut and entered total snowball; on Hoth, we see how the rebels have to close the shield doors to their base at night to keep out the cold, how their tauntauns – the bipedal creatures that the rebels ride that are adapted to extreme weather – die from the cold, and how even their snowspeeders have trouble adapting to the freezing conditions. When Luke and Han Solo are trapped outside at night, they are not expected to survive (yet somehow, they do, partly thanks to the injured and frostbitten Luke being stuffed inside an open wound in the belly of a dead tauntaun).

## Life on the ice

However cold it is, what's left of life on Hoth is clearly able to linger, with some form of food chain left for an apex predator like the wampa to subsist on. Admittedly, perhaps not too much thought was placed into developing the ecology of Hoth; more care was taken by Le Guin when describing Gethen. On her planet of always-winter, the largest land animals, such as the pesthry, are herbivores no larger than foxes, and the people living there eat mostly fish and vegetables.

We can look to our own Antarctica – Earth's closest environment to fictional winter worlds such as Hoth and Gethen – to see how limited life is in such conditions. Ninety-nine per cent of the continent is permanently covered in ice – well, at least until climate warming melts it – and the ice-bound biosphere is pretty shallow: the largest species living on that ice is only 6 millimetres (around ¼ in.) in length, and it has to deal with freezing temperatures that can drop as low as −80 degrees Celsius in the midst of winter, six months of darkness, harsh winds and dry conditions (yes, dry – Antarctica is the largest desert on Earth, with an average of 10 millimetres (⅖ in.) precipitation per year). Life fares a little better in the sparse areas where bare rock is exposed, home to 'micro-forests' of lichen and moss, hardy grass and the occasional flowering plant, fed by summer meltwater and inhabited by various microscopic insects such as tardigrades.

With all that in mind, it's hard to see how wampas would be able to survive on Hoth for very long.

In the Southern Ocean around Antarctica, things are very different. There's a well-supplied food chain, starting with ocean-borne phytoplankton that form enormous algal blooms, large enough to be seen from space, in the water. Phytoplankton is grazed upon by the trillions of small crustaceans called krill that inhabit the Southern Ocean, and a blue whale can easily gobble up a million krill every day. Fish, penguins and seals all form various links of the chain, and the total biomass is staggering – the krill alone make up the largest biomass of any wild animal on the planet. It is clear that if life is to survive a snowball stage, it needs at least some oceans and seas to remain habitable.

One way that is possible is through huge lakes that persist beneath ice sheets. Antarctica is home to over three hundred ice sheets where the frozen land meets the Southern Ocean, providing white ceilings over bays, inlets and lakes that shelter habitats deep beneath them. In some cases those habitats have been enclosed by the ice and cut off from the outside world for millions of years. The most famous example is Lake Vostok, a huge freshwater lake that is the largest of Antarctica's subterranean lakes, buried 3.7 kilometres (2½ mi.) below the ice and extending to 250 kilometres (155 mi.) in length, 50 kilometres (30 mi.) in width and half a kilometre (3⁄10 mi.) in depth. Oxygen dissolves into the lake from the ice above; indeed, the lake is supersaturated with oxygen, containing an abundance fifty times greater than typical freshwater lakes elsewhere on Earth. When boreholes drilled into the lake are opened, pressure is released and the water gushes upwards, literally fizzing like opening a bottle of pop.

To be clear, scientists have not yet confirmed the existence of life in Lake Vostok, but microbial life has been discovered in the layers of ice that have formed over the lake, as scientists drill cores into the ice. Further research is being complicated by the need to keep the pristine environment of Lake Vostok clean of outside contamination, particularly from bacteria on the drilling equipment and the mixing of antifreezes such as kerosene used to prevent the borehole freezing over. However, Lake Vostok is potentially a significant habitat where life may have been evolving on its own, cut off from the rest of the planet, for at least 15 million years. That life will be a kind of 'extremophile', adapted to thrive in the high oxygen concentrations that might kill unadapted life and, if nothing else, the ecosystem within Lake Vostok could provide a unique experiment in independent evolution. Life has reached every other nook and cranny of our planet, so why not Lake Vostok too?

We also see something similar on the icy moons of the giant planets in our solar system. As discussed in the previous chapter about ocean worlds, Jupiter's moons Europa and Ganymede, and Saturn's moons Enceladus and Titan, among others, are very much like planetary-scaled Lake Vostoks – potentially habitable oceans submerged and cut off by dozens of kilometres of ice.

If life is to survive long-term on a snowball planet, then subglacial lakes are one way it could do so. We don't know if there are any of these dark oases on Hoth, but we're much more likely to find thriving life in the waters beneath the ice on snowball worlds than on the exposed surface.

## Melting the snowball

Would Hoth and Gethen remain snowball planets for ever? In the case of Hoth, the answer is yes, figuratively, our glimpse of it in *The Empire Strikes Back* seared into cinematic history. When we think of *Star Wars*, and of Hoth, we will always think of that ball of ice.

Le Guin put more thought into Gethen. She saw it as an evolving world, one that was perhaps different in the past, and will perhaps be different again in the future. In doing so, she hit upon the explanation for how Earth exited its snowball states.

A snowball state is a serious affair, but it need not be permanent. The snowfall that builds up the ice sheets and marauding glaciers that expand across the surface removes water vapour and carbon dioxide from the atmosphere, both of which are greenhouse gases. Whether it is through this snowy precipitation, or via the ingestion of carbon dioxide by life, ultimately the atmosphere becomes depleted in these vital gases, and in order to bring about the end of the snowball they must be replenished. The most effective means of doing so is volcanoes belching out carbon-based greenhouse gases.

In *The Left Hand of Darkness*, Le Guin alludes to how Gethen has gradually seen its volcanic activity increasing over the past 10,000 to 20,000 years, and how the carbon dioxide entering the atmosphere from the volcanic plumes presages an end to the ice age and an increase in the average global temperature of the planet to about 22 degrees Celsius, a marked improvement on the frigid 5 degrees Celsius or so that the thermometers reach in the story. This is exactly how Earth has thawed out in the past.

'We think that the reason a planet escapes its snowball state eventually is that when it is cold, weathering is slow and carbon-dioxide removal from the atmosphere is at a very slow rate, so the volcanoes can outpace it,' says Paradise.

Earth has its own in-built thermostat, the carbon–silicate cycle, of which volcanoes are an essential cog. The cycle begins in the atmosphere when the climate is warm; such temperatures bring with them the tendency for plenty of rain because the water content of the atmosphere is high due to evaporated moisture. This rain carries with it dissolved carbon dioxide, which results in two consequences: it removes a potent greenhouse gas from the atmosphere, thereby preventing Earth's climate from running away with itself and growing too warm; and on the planet's surface the carbon-rich rainwater reacts with silicate rocks, weathering and eroding them. In Chapter Four, we learned how this carbon makes its way into life that eventually perishes and sinks to the sea floor where, at tectonic fault lines such as the Mid-Atlantic Ridge, subduction drags it into Earth's mantle. For millions of years, it becomes mixed within the mantle as the temperatures above ground cool in response to the sequestration of the carbon dioxide, as the carbon awaits the day that it is spat back out as carbon-dioxide emissions from volcanoes. If enough carbon dioxide builds back up again in the atmosphere, it resets Earth's thermostat, the ice melts and the planet re-enters an interglacial state. 'That's why volcanoes are important, because they replenish carbon dioxide,' says Paradise.

Over geological timescales, Earth would experience warmer and cooler periods simply because of the carbon–silicate cycle acting like a thermostat. Throw into the mix the Milankovitch cycles, and the risk of tipping into a snowball state, and the carbon–silicate cycle becomes more than just a planetary thermostat oscillating between slightly warm and slightly chilly; instead, it acts as a safety mechanism against extreme natural climate change that can threaten the habitability of Earth. (It should be pointed out that current human-induced climate change is circumventing this cycle, and nature won't solve the problem for us.) Volcanoes and tectonic activity are a hazard to any life nearby in the present day, but in the long term they are why life still persists on our planet.

Herein lies the potential problem. Planetary scientists do not know how common plate tectonics is in the universe. However, plate tectonics need not be a prerequisite for volcanism, and indeed, Earth is not the only planetary body in the solar system to have volcanoes. We see the enormous but extinct volcanoes of Mars, including the giant Olympus Mons, so tall at 21.9 kilometres (13½ mi.) that its peak pushes out of the thin Martian atmosphere, while NASA's various Mars rovers have found plentiful evidence for ancient volcanic activity on the Red Planet billions of years ago. A more recent development is the discovery of active volcanoes on Venus by scientists searching old imagery from NASA's Magellan mission, which orbited Venus during the 1990s. We're not talking violent pyroclastic flows from explosive volcanism; instead, the scientists, led by Robert Herrick of the University of Alaska Fairbanks, found changes in the shape of a Venusian volcanic vent. This is despite Venus not having plate tectonics. Their absence is because Venus is the hottest (with surface temperatures averaging 475 degrees Celsius) and driest planet in the solar system, and water is required to lubricate tectonic plates so that they can ride the mantle, move over one another and subduct. Planetary scientists describe Venus as having a 'stagnant lid', whereby the crust of a planet is solid, and volcanism comes from convection currents in the mantle that cause plumes of magma to push up against the underside of the crust and eventually break through. Though carbon dioxide can be pumped into the atmosphere by these volcanoes, there's no rain to remove it again, so the temperature just keeps increasing as more carbon dioxide enters the atmosphere. The planetary thermostat breaks, leaving Venus to evolve into a world befitting Dante's *Inferno*, quite the opposite of a snowball planet.

As we learned in Chapter Four, there's significant debate in the exoplanet community about whether the rocky worlds that astronomers are now finding en masse around other stars can also support plate tectonics, or if they have stagnant lids too. One of the most common forms of rocky exoplanets being discovered is the super-Earth – chunky, rocky worlds with masses up to ten times greater than our planet. On these worlds, the problem isn't necessarily high temperatures that cause water to evaporate and then escape into space, as on Venus; it could be that their large masses result in crusts too thick to break into plates that can slip and slide over one another, and therefore allow material to enter the mantle. It's not even clear if such worlds could have the kind of hot-spot volcanism that Venus has, and Mars had (and which formed the Hawaiian chain of islands on Earth), because in such cases the crust still has to be pierced by an upwelling plume of magma from the mantle. Without that two-way conveyance of carbon, the planetary thermostat doesn't work, and a planet that might otherwise have been habitable, based on its location in the habitable zone, slips into one of two extremes: a Venusian hot-house or a Hoth-style snowball.

'If you have a planet that doesn't have plate tectonics, and doesn't have much volcanism, can the carbon dioxide still escape from the mantle?' wonders Paradise. 'You might end up with a planet where all the carbon dioxide gets locked into the mantle, and volcanism shuts off and you end up with a runaway snowball that might suppress volcanism – we don't fully understand the feedback between surface temperature and volcanism all that well. In that case, the snowball would become permanent, at least until the star becomes brighter and melts it.'

The ability of a planet to be frozen or temperate at different times is called climate bi-stability, says Howe. You could even throw in a third state, a Venusian greenhouse, making it more tri-stability. 'All three of those could exist at Earth's distance depending upon the history of the planet,' says Howe. 'But states in between those are unstable and will drift towards one of the stable states.'

It's remarkable that a planet could exist in any one of these three states while sat in the habitable zone, which just goes to show that the habitable zone as a concept is severely overly simplified. And given climate bi-stability, we might also ask: just how common are snowballs, even among Earth-like worlds? While Earth has frozen over several times in its history, the amount of time spent as a snowball is actually a pretty small fraction of Earth's total history. Do all Earth-like planets in the habitable zone have to endure a snowball state at least once in their lifetime?

'It's a fascinating question,' says Paradise. 'There's probably a slice of the habitable zone where you would expect to find snowball planets if an Earth-like atmospheric chemistry is normal, so assuming that Earth is not unusual, then snowball planets might be pretty common.'

Once again, science fiction has been able to touch on a few truths about the nature of planets. Yes, it's another example of the trope of taking a particular environment from Earth and expanding it to a global scale, this time an Antarctic environment. Yet, just like the desert-planet trope and the ocean-planet trope, the snow-planet trope has a solid basis in fact. Hoth, Gethen, Darkover and all those other snowy worlds seen in science fiction are not just fictional exoplanets – they could easily have been Earth, once upon a time.

# 6

# WORLDS OF ALWAYS-DAY AND ALWAYS-NIGHT

When Paul McAuley was searching for exoplanets to fill one of his fictional universes, he chose worlds orbiting the most populous type of star in the cosmos. In his novel *Something Coming Through* (2015), the first in a duology set in a universe where enigmatic aliens known as the Jackaroo contact early twenty-first-century Earth by declaring that they are 'here to help', much of the action is set on the world of Mangala. The name is the Sanskrit word for Mars, but though it bears some similarities, Mangala is not Mars. Rather, it's a relatively dry desert world orbiting a red dwarf star 20,000 light years away. Mangala is a gift from the Jackaroo, one of fifteen such 'gift worlds' orbiting red dwarf stars and reachable via shuttles that travel through artificial wormholes. The human settlers on these planets find the ruins of other, ancient civilizations – the remnants of previous client species of the Jackaroo. What became of them? What will become of humanity? And what do the Jackaroo want?

We discover some of the answers in the sequel novel, *Into Everywhere* (2016), in which McAuley takes us on a tour of many more worlds, mostly planets around red dwarfs ranging from temperate lands to mini-Neptunes, and a few ocean worlds thrown in for good measure.

Why red dwarfs? Well, let's think about the Sun's neighbours. In 2018 a team of astronomers called RECONS (REsearch Consortium On Nearby Stars), led by Todd Henry of Georgia

State University in the United States, compiled a census of everything known within a radius of a little over 30 light years from the Sun. They counted 378 stars, 50 brown dwarfs (an object not quite massive enough to generate the interior temperatures required for the nuclear fusion of hydrogen in its core, the hallmark of a star) and more than 50 planets (a number that continues to climb, having reached over 250 planets at the time of writing in June 2024). Of the 378 stars, 284 are red dwarfs. In other words, three out of every four stars are red dwarfs, a type of star far smaller and cooler than our Sun. No wonder the Jackaroo's gift worlds are all orbiting red dwarfs. The damn things are ubiquitous, which may be great – or terrible – for the potential of other habitable planets in the universe. As we shall see, it really could go either way, but while science fiction likes to take a more optimistic view, is it the realistic view?

## Planetary systems in miniature

There are two things that seem obvious about red dwarfs: they are red, and they are small. Their red colour is a result of their being cool, or at least cool for a star. A red dwarf's surface temperature (that is, how hot the visible photosphere is) languishes south of 3,800 degrees Celsius and can be as low as 2,300 degrees Celsius, which, let's face it, is still scorching hot. Compare it to our life-giving Sun, however, which has a photospheric temperature of 5,500 degrees Celsius. And those massive blue-white stars that end their lives in a bang as supernovae? Their surface temperature can reach 30,000 degrees Celsius. So yeah, as stars go, red dwarfs are lukewarm. Sometimes you might hear astronomers refer to them as 'M-dwarfs'; this astronomical jargon hails from the popular mnemonic 'Oh Be A Fine Girl/Guy, Kiss Me', which is a memory aid for the early twentieth-century Harvard astronomer Annie Jump Cannon's classification scheme for stars based on their luminosity and colour/temperature. She used the letters O, B, A, F, G, K and M to denote the hottest, brightest stars to the coolest, dimmest stars – hence why the coolest red dwarfs are also M-dwarfs.

They're also small, the most diminutive having a mass just 8 per cent that of our Sun and a correspondingly tiny diameter. The largest red dwarfs, however, have nearly two-thirds the mass of our Sun. Their modest stature is carried over into their planetary systems, which tend to be filled with small, rocky planets huddled closely around their parent star and nary a gas giant to be found, with only 2 per cent of red dwarfs having accompanying Jupiter-like planets. They just don't have enough raw materials available to build such giant worlds often.

Even in planetary systems orbiting red dwarfs, we're still describing distances between planets in terms of millions of kilometres. These are huge distances in human terms, no doubt, but to really appreciate how close the planets and their star truly are, let's think about the TRAPPIST-1 system, which is a real-life system of one red dwarf orbited by no less than seven rocky planets, three of which are in the habitable zone, and all located 40 light years away from us.

The seven planets are 'so close to one another it's ridiculous', according to Amaury Triaud, an astronomer from the University of Birmingham and one of the co-discoverers of the TRAPPIST-1 system. Triaud is so invested in the belief that red-dwarf planetary systems are worthy of study that he manages the TRAPPIST-1 website (www.trappist.one), a public page collating everything we've learned about these seven planets so far. When his team's discovery paper was published in *Nature*, Triaud and the journal's editors even commissioned Triaud's friend, the Swiss SF author Laurence Suhner, to write a short story about TRAPPIST-1 – 'One of the first stories about a planet orbiting an M-dwarf,' says Triaud.

In Suhner's story, named 'The Terminator' (no, not that one), a settler on TRAPPIST-1e describes how she looks up to see a neighbouring planet, TRAPPIST-1f, in the sky. At a distance of 1.4 million kilometres, planet f is so close that it appears as large as the Moon does in Earth's sky. Earth's Moon averages a distance of only 384,400 kilometres (238,855 mi.) from us, but TRAPPIST-1f would appear just as big in the sky because it is 9,855 kilometres (6,125 mi) larger than the Moon (and is even 858 kilometres (533 mi.) larger than Earth). Some of TRAPPIST-1's worlds are even more tightly packed; the distance between TRAPPIST-1b and -1c is just 640,000 kilometres (397,680 mi.) – less than twice the distance between Earth and our Moon – when the two planets are both in line on the same side of their sun. And both worlds are so close to their star that one 'year' on them is equal to one-and-a-half Earth days and two Earth days and ten hours, respectively. Even TRAPPIST-1e, sat in the middle of the habitable zone, completes one orbit every six days and two hours at a distance of 4.3 million kilometres. It's able to potentially be habitable because its star is much cooler than the Sun, so the habitable zone is much closer in. And to really boggle your mind, the outermost planet, TRAPPIST-1h, is just 9.3 million kilometres from the star. The entire seven-world system is packed into a volume less than one-sixth of the distance between our Sun and Mercury, the solar system's innermost planet.

'If you were in your back garden with a telescope on one of these planets, you'd be able to actually see a city on one of the other planets,' says Triaud. When planets b and c are close to one

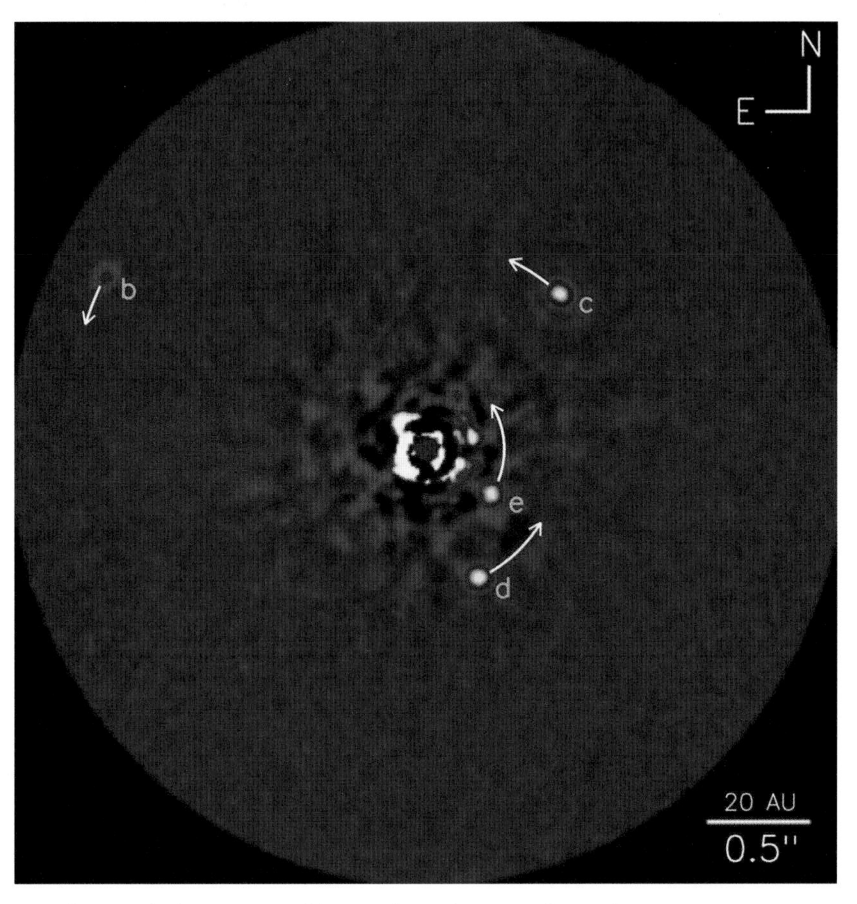


An actual, direct image of four real exoplanets in the HR 8799 system, which is 133 light years away. Because they are so far away, we see the planets as only points of light. The star at the centre of the system has been blotted out by a coronagraph. The arrows indicate the planets' direction of motion as they orbit their star.

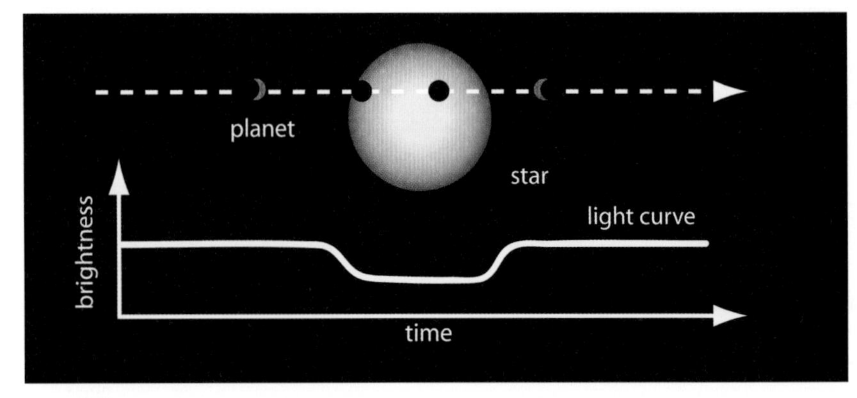


Astronomers are able to detect an exoplanet when it transits its star by the way in which the planet blocks some of the star's light. The bigger the planet, the more light it blocks.

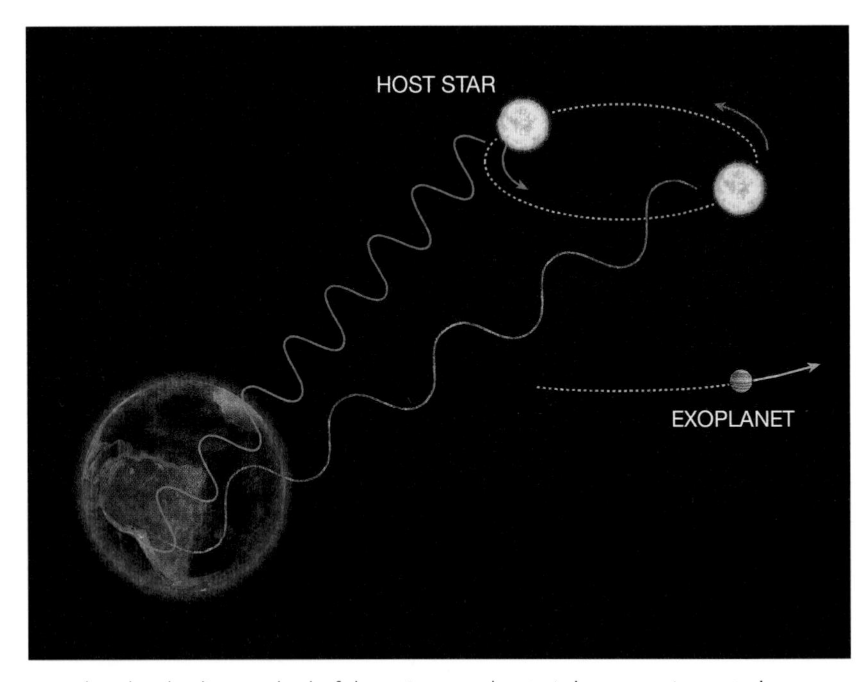


Another leading method of detecting exoplanets is by measuring a star's radial velocity. This is the changing Doppler shift of its light as the star subtly moves towards and away from us, at speeds of just metres per second, while revolving around a common centre of mass shared with an orbiting planet, or planets. The bigger an exoplanet and the closer it orbits to its star, the stronger the star's Doppler shift.

Swirls of water-ice at Mars's north pole. In the planet's distant past, when it was warmer, this ice would have run as liquid through rivers and in oceans.

Artist's impression of what it might be like to stand on the surface of Proxima b. Because Proxima b is tidally locked, its red dwarf sun would not appear to move in the sky from any given location. For the star, Proxima Centauri, to appear this low means that this artwork is depicting a region near the planet's day–night terminator.

The habitable zone in the TRAPPIST-1 system. The planets closest to the star are too hot for liquid water, hence the steam. The outermost planets are too cold, hence the ice. However, two, and perhaps three of TRAPPIST-1's planets – namely planets e, f and g – could be at just the right distance for water to be liquid on their surface.

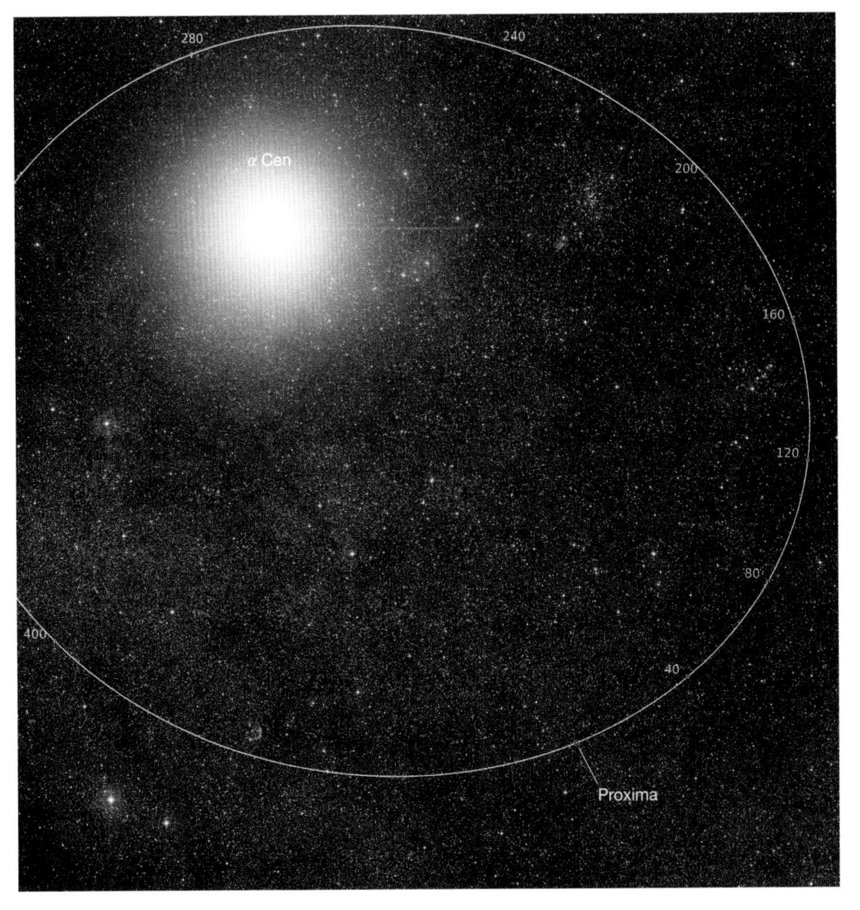


The alpha and Proxima Centauri system. In this real image of the night sky, alpha Centauri a and b are too close to one another to be separated. The red dwarf Proxima Centauri is on a long ~550,000-year circumbinary orbit around the double alpha Centauri system (the graduations on the orbit are in thousands of years).

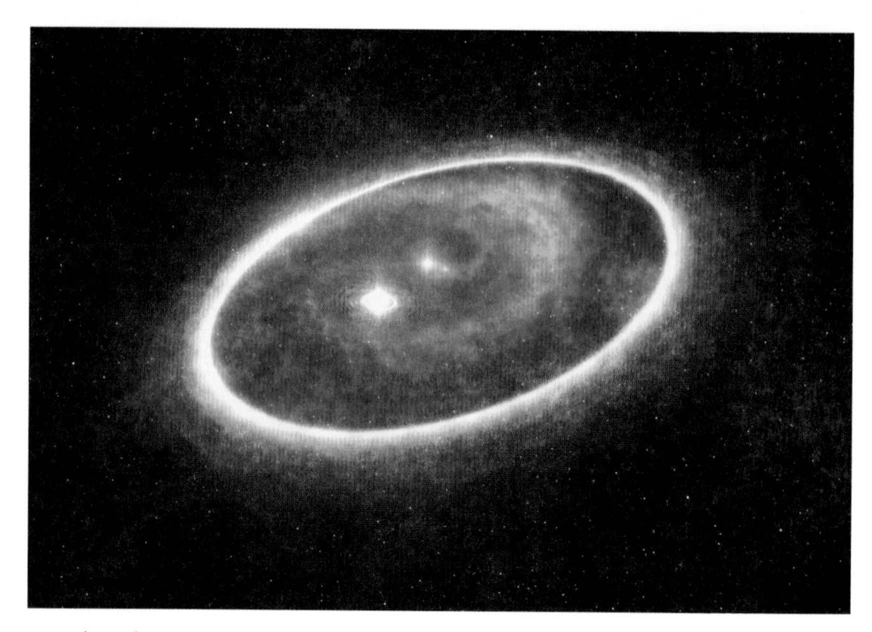

Artist's impression of the planet-forming disc around the young triple star system GG Tauri A. Orbiting around this three-star system and disc are another two stars (not pictured here) bringing the total number of stars in the system to five.

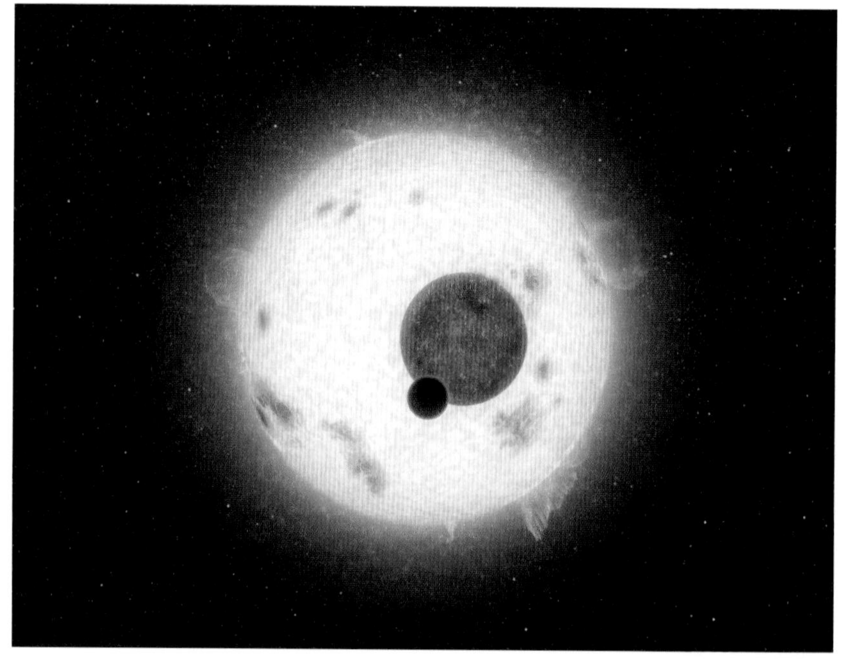

The double-star system Kepler-16 and its circumbinary planet Kepler-16b (seen in silhouette) are shown in this artwork transiting one another.

Artist's impression of an exoplanet orbiting a pulsar. The planet is experiencing an aurora over its magnetic pole as a sleet of charged particles are caught by the planet's magnetic field and interact with the atoms and molecules in its tenuous atmosphere, causing them to glow.

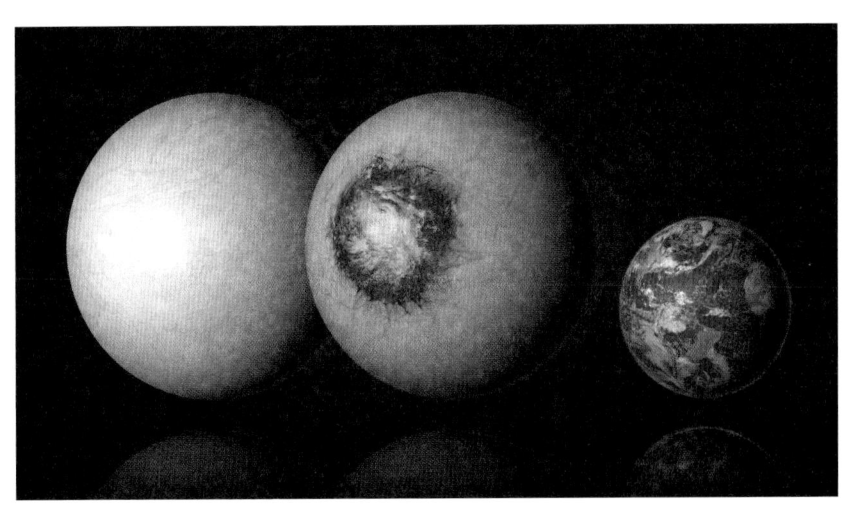

Artist's depiction of LHS 1140b, a planet 39 light years away from Earth, shown to scale on the right. Tidally locked worlds on the edge of the habitable zone might be totally frozen, or potentially sport a temperate region at the sub-stellar point, giving the impression of an eyeball planet.

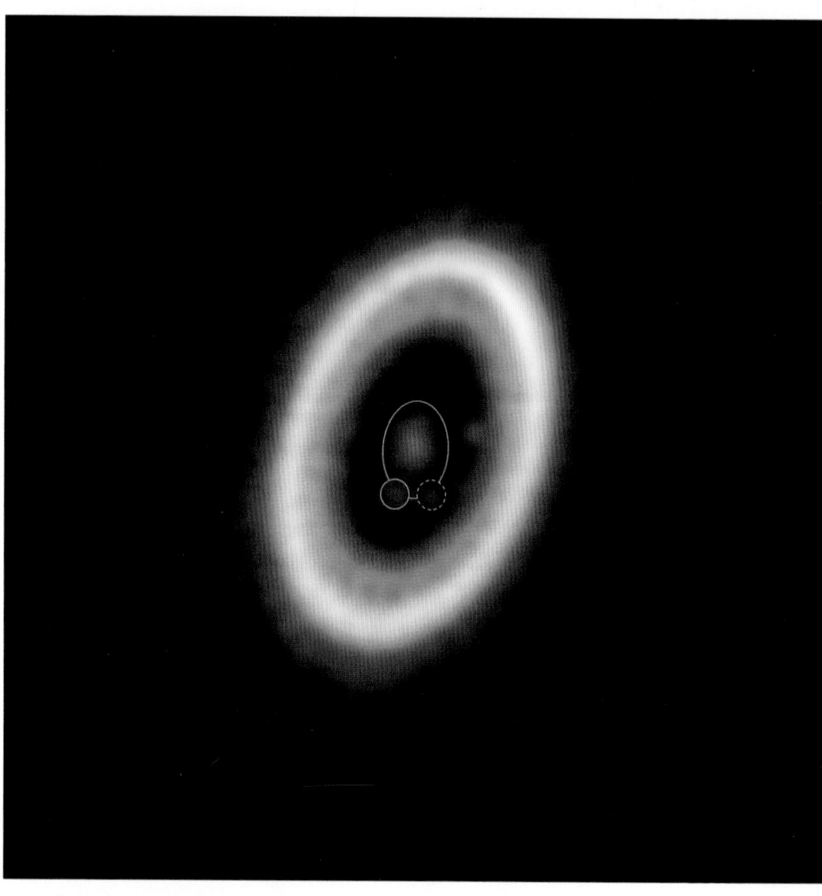

A real image of the PDS 70 system, taken by the Atacama Large Millimeter/submillimeter Array (ALMA). We can see a dusty debris disc, as well as one of the system's planets – PDS 70b, highlighted by the solid circle – and another object, seemingly a trojan planet sharing PDS 70b's orbit (dashed circle).

another, the tidal forces can be three times stronger than those experienced on Earth, so any interstellar surfers might want to head to them for the waves – that is, if they have oceans. And the most science-fiction-like concept is that travel between these planets would be no more difficult than the Apollo missions to the Moon.

## No boom today

The coolest thing about planets around red dwarf stars is that they are on the bleeding edge of astronomers' quest to find other habitable planets. Not that this was always the case. When that quest began in the 1990s the goal was to discover habitable worlds around Sun-like stars, because that's all we knew. Over time, however, as more and more rocky worlds have been discovered orbiting red dwarfs, this has changed; today, red-dwarf planets such as the worlds of TRAPPIST-1, Teegarden's Star b (first mentioned in Chapter Two as the exoplanet with the highest Earth Similarity Index rating) and even the planet Proxima b, which orbits the closest star to the Sun – namely Proxima Centauri, located just 4.2 light years away – are at the forefront of our search for habitable worlds. Now science fact is beginning to bleed into science fiction.

'I was convinced a few years back that there was going to be a flood of novels set on planets around red dwarfs because they are just such a rich setting,' says Charlie Jane Anders, whose books include *The City in the Middle of the Night* (2019), set on just such a world. 'I read up on these planets and it was obvious they would be where we would settle if we left our solar system.'

However, Anders feels that SF authors in general have been a little slow in taking advantage of the real-life discoveries that were happening around them. 'I thought there was going to be a huge boom in novels about these planets,' she says. 'But there really hasn't been and I'm very surprised.'

There is perhaps a couple of reasons behind this. One might be that it's still cutting-edge astronomy and might not have percolated fully into the SF community, or the readership in general. Another reason might be cultural, notes St Andrews's Emma Puranen, who if we remember is collating planets from fiction. 'I've noticed a decrease in stories with an established non-native human population since 1995, which might be related to increased awareness of the harms of colonization, for example,' she says.

Red-dwarf planets provide a new frontier for exploration and human settlement, should we ever become an interstellar civilization, but our growing awareness of the damage done by colonial powers in the past on Earth has made SF authors wary of celebrating colonialism on other planets. Indeed, Anders's *The City in the Middle of the Night* touches on how colonization can affect native inhabitants. In McAuley's Jackaroo books, the roles are reversed: the gift worlds have no current native inhabitants, bar wildlife, but the Jackaroo are older and more technologically developed than humans, and it is humans that are on the receiving end of dealing with the consequences of having been contacted. Meanwhile, in Becky Chambers's *To Be Taught, if Fortunate* (2019), the human astronauts are on a scientific mission, being careful not to harm the environments and the native life of the red-dwarf worlds that they explore, but when one indigenous life-form comes to accidental harm, the characters are heartbroken.

## Ribbon worlds

There's perhaps a more practical reason why red-dwarf planets have not necessarily become the great frontier in SF that Anders predicted. For all the promise that they hold, planets orbiting red dwarfs are almost certainly not going to be just like Earth, and in the worst-case scenario they could be rendered completely inhospitable.

'Most of these planets will be tidally locked,' says Abel Méndez: 'Not necessarily all, but probably most.' When we say that a planet is tidally locked, we mean that it always shows the same face to its star: one hemisphere is in constant daylight, and the other is in perpetual darkness. This happens when a planet is very close to its star. The planet is still rotating, but the gravity of the star is actually able to subtly distort the shape of it, forming a tidal bulge on the planet itself. It's a more extreme version of the way that the Moon and the Sun tug on the oceans to form the tides on Earth. A planet with a tidal bulge doesn't necessarily become egg-shaped, but if you measured the diameter of the planet pole-to-pole, and then west to east, you'd find enough difference for the star's gravity to lock on to the excess bulge. In doing so, a star can force any planets at close quarters to rotate such that they always show their bulge to the star. To achieve this, a planet must complete one rotation on its axis for every one orbit of its star; such a planet's day and year become the same length, and astronomers call this phenomenon 'tidal locking'.

Sometimes, the concept of tidal locking gets misunderstood. In the SyFy TV series *The Ark*, which debuted in 2023, Earth is dying, and a fleet of spacecraft known as Arks set sail for the stars, carrying human settlers to new worlds. When they arrive at Proxima b, the crew of the Ark One and a renegade ship, Ark 15, seem surprised to discover that Proxima b is tidally locked – something you'd expect them to already know – but furthermore, the show's writers think this means that Proxima b doesn't rotate at all. In reality, this is not the case: tidally locked worlds do rotate, it's just that their rotation is synchronous with their orbital period.

In *The Ark*, one crew-member calls Proxima b an 'eyeball planet', because most of the surface is frozen apart from a region around the sub-stellar point, which is too hot. The geological arrangement of a circle of land amid white ice gives the impression of an eyeball. They then attempt to kick-start the planet's rotation by deploying powerful neodymium magnets in a ring around the planet to generate a magnetic field strong enough to put pressure on Proxima b's magnetic poles and make it start spinning. Something goes wrong with the experiment, however, and the entire planet is destroyed. Yes, the science behind it is arrant nonsense, but *The Ark* is one of the first times that a tidally locked world around a red dwarf has been presented in live-action SF.

During the summer of 2024, astronomers using the JWST claimed to have possibly found an eyeball planet for real. LHS 1140b is a super-Earth orbiting in the habitable zone of a red dwarf star 48 light years away. The JWST observed LHS 1140b transit its star, and tentatively detected nitrogen in its atmosphere. This suggests that LHS 1140b may have an atmosphere similar to Earth's, which is 78 per cent nitrogen, rather than an inhospitable hydrogen envelope. Furthermore, the planet's density is low enough to suggest that up to 20 per cent of its mass consists of water. Global circulation models for this tidally locked world, which is 1.7 times bigger than Earth, point to its night-side being completely frozen, but on the dayside, beneath the substellar point, the ice would melt to form a 4,000-kilometre-wide (2,485 mi.) ocean with cosy temperatures of 20 degrees Celsius.

A circular blue ocean amid a mass of white ice – if the findings are correct, then it would be a true eyeball planet.

It's not just red-dwarf planets that become tidally locked; hot Jupiters around Sun-like stars have been observed to be tidally locked too. In a slightly different scenario, the Moon is tidally locked to Earth, hence we always see the same face of the Man in the Moon (though it's not quite the same scenario because the far side of the Moon isn't always dark; it receives just as much sunlight as the nearside). Little Mercury was also once thought to be tidally locked to the Sun, and we can see this in the science fiction of the 1940s, '50s and early '60s, one example being Isaac Asimov's 1956 novel *Lucky Starr and the Big Sun of Mercury*. Asimov's *Lucky Starr* series was a set of space-adventure novels for young adults ('juveniles', or 'juvies', as such works were described at the time) that are great fun, if one makes allowances for the era they're written in. Anyway, *Big Sun of Mercury* is, as you no doubt guessed, set on Mercury, but a Mercury that is tidally locked, and the story takes place in a temperate zone that forms a strip on the terminator between always-day and always-night. Another story of this era set on a tidally locked Mercury is Larry Niven's first published short story, 'The Coldest Place' (1964), where the eponymous 'coldest place' is the fictional dark side of Mercury.

Asimov called such worlds 'ribbon worlds', because it was assumed that the dayside would be too hot for life, and the nightside too cold, so the only habitable region would be a narrow ribbon that ran from pole to pole around the day–night terminator.

Tidally locked worlds continued to appear sporadically in SF prior to the discovery of exoplanets – an episode of *Star Trek: The Next Generation* (a season-two episode from 1989 entitled 'The

Dauphin'), for example, featured the *Enterprise* transporting a young politician from a world called Daled IV that is tidally locked. We don't get to see her planet, but Lt Commander Data describes Daled IV as rotating once per revolution, such that one side is constantly dark and the other constantly light. Data ascribes the constant warfare on Daled IV as being the result of the opposite hemispheres evolving two different and incompatible cultures.

Today we know that Mercury is not tidally locked. Rather, it is in a 3:2 resonance, whereby it rotates on its axis three times for every two orbits of the Sun. This is what Méndez meant when he said that 'not necessarily all' red-dwarf planets will be tidally locked. Mangala, in *Something Coming Through*, shares a similar resonance to Mercury, as do a few other red-dwarf planets in McAuley's novels, spun up a long time ago by magical alien technology. This 3:2 resonance means that both Mercury and Mangala experience sunrises and sunsets, and on Mangala possibly even brief seasons, with the changing weather that different seasons can bring. A classic example in the novel is that Mangala, being a desert world, is subject to dust storms that can last several weeks during the planet's 'summer'.

True tidally locked worlds can't have seasons because they are forced to straighten up and rotate without being tilted relative to the ecliptic plane – the plane in which the planets orbit around their star. Earth has an obliquity of 23.4 degrees, and this means that we have seasons, because the intensity of solar heating incident on different parts of the planet varies throughout the year as a result of this tilt. And on Earth, we are familiar with life's cycle throughout the seasons – lambs and chicks in the spring, the blossoming of trees into the summer, falling leaves in the autumn and animals scurrying into hibernation for the winter. On a tidally locked planet with no tilt, this familiar cycle of life could not occur. You would only find different conditions at different locations on the planet, depending on the height of the stationary sun in the sky. 'You'd instead have to walk to see a different season, which sounds nice,' says Méndez.

This highlights the greatest consequence of tidal locking: the fact that the sun would never appear to move in the sky from any given location. It just hangs, like some omnipotent overlord, blazing down on the sub-stellar point on the permanent dayside. If you wanted to see the sun higher or lower in the sky, you'd have to trek across the land to a different location. (But you'd at least know your latitude and longitude based on where the sun was in the sky.)

Between the dayside and nightside, there's an intermediate zone, the ribbon that Asimov described, where the sun sits on the horizon, casting the land in never-ending twilight. From the days of Asimov writing *The Big Sun of Mercury* until recent times, it had been thought that this ribbon would be the only potentially habitable land on a tidally locked planet. Indeed, Charlie Jane Anders set the action in *The City in the Middle of the Night* within this ribbon.

'I found it fascinating that on a tidally locked planet people would only be able to live in the twilight area,' she says, joking that she wanted to name her novel *The Twilight Zone*. 'It wasn't until much later, when it was too late to change the novel, that I talked to scientists who said the new thinking is that people could live elsewhere on the planet, but by then I'd already made my bed.'

Not that Anders would have been persuaded to change it even if she'd had time: 'I just loved the image of day and night being over here and over there, and I just thought that felt very poetic and I didn't want to change that.'

So, what caused planetary scientists to change their minds? The habitability of a tidally locked planet depends upon the planet's ability to efficiently redistribute the searing heat that is incident upon the dayside by transporting it around to the hemisphere of never-ending night, so that the dayside doesn't become too hot, and the nightside doesn't become too cold. A planet can facilitate this via large-scale circulation within its atmosphere.

Global circulation models, which are systems devised to describe how an atmosphere operates, are able to predict what we might find in the atmosphere of a planet orbiting a red dwarf.

'We expect that for tidally locked planets there will be a big convection cell,' says Harvard's Robin Wordsworth. This cell will see rising air on the dayside, focused at 'high noon' – the substellar point, which is the part of the planet at the equator where the sun is directly overhead. Wordsworth adds, 'The rising air travels all the way around to the nightside of the planet and comes back again, and you can have sub-tropical regions at the mid-latitudes away from the sub-stellar point.'

The circulation would keep things mostly balmy, rather than incessantly, suffocatingly hot, although the sub-stellar point probably wouldn't be much fun, like forever standing in the desert at mid-afternoon. But – there's always a but – this all depends on whether a tidally locked planet orbiting a red dwarf can keep hold of its atmosphere. Red dwarfs, it turns out, have a habit of blowing away the atmosphere from their planets.

## In the line of fire

Despite their small stature, red dwarfs are remarkably violent, unleashing angry flares of radiation for the first few billion years of their lives that can strip a planet of its atmosphere.

'M-dwarfs are incredibly high-energy emitters, especially early in their lifetimes,' says Wordsworth: 'The first billion years for a planet around an M-dwarf is rough, and they have to survive that in order to have a chance of being habitable later down the line.'

Even the more mature red dwarfs are prone to cataclysmic outbursts of radiation; in 2019 astronomers measured a flare erupting from Proxima Centauri, a star nearly 5 billion years old, that saw it increase its brightness in ultraviolet light by a factor of 10,000. The flare itself was one hundred times more powerful than any similar flare seen on our Sun, and although it lasted just seven seconds, all that ultraviolet radiation would have been extremely hazardous to potential life on the orbiting rocky planet Proxima b. Even those more moderate red dwarfs experience frequent smaller flares, as observed during a ten-year study by NASA's Galaxy Evolution Explorer (GALEX) spacecraft. The radiation exposure is incremental, and potentially deadly.

Yet there is hope that some planets may be able to resist this deadly onslaught in much the same way that Earth wards off the calmer solar breezes of our Sun's radiation and the occasional gust of a coronal mass ejection, when the Sun 'burps' out a huge cloud of ionizing particles. Our planet is protected by the defensive shield that it wields, a magnetic field that can deflect the charged particles in the solar wind. It would take quite an impressively strong field to ward off the torrent of radiation close to a red dwarf, but it's not beyond the bounds of possibility. Not that all astronomers are worried about the radiation anyway.

'As long as we still can't say what conditions are required for the origin of life, then I don't give a damn about the activity of the star,' say Triaud, who is highlighting the fact that there could be other possibilities for life – perhaps living underground or within oceans beneath a frozen crust, or maybe life on a planet orbiting a red dwarf has evolved resistance to radiation like some extremophile microbes on Earth. The flares may strip the atmosphere, but that's not necessarily a problem for biology if we don't know how varied and hardy alien biology can be.

'Too many people assume that these planets are going to be no good for life,' says Triaud. 'I say, let's go and observe them.' Indeed, this is exactly what is currently happening, with our hopes for the seven planets of TRAPPIST-1 in particular being put to the test by the JWST, which is searching for atmospheres wrapped around those planets. Its initial results have not been promising – the two innermost planets, b and c, do not seem to have sufficiently thick atmospheres and are quite possibly airless. Now, these are too close to their star and too hot to be habitable to life as we know it anyway, but it doesn't bode well. Perhaps by the time you read this, we'll have more definitive answers regarding TRAPPIST-1's planets, including the three worlds that lie within its nominal habitable zone. But it doesn't look good. 'The first place we've been to look, and it seems like things didn't go well,' says Wordsworth. 'Their atmospheres got stripped away by their host star early on.'

The problem of flares is rarely mentioned in SF, although they do pose a danger to the characters in one of the first novels to feature a tidally locked planet around a red dwarf, Stephen Baxter's *Proxima* (2013). It is named after Proxima Centauri, but in an act of foresight, it was written three years before Proxima b was discovered. Baxter chose to put a planet around Proxima Centauri 'just because Proxima is the nearest star . . . It seemed to me that because red dwarfs are the most common type of star, we might discover a planet around a very near star. I remember receiving a phone call to say they'd found my planet after they discovered Proxima b!'

In Baxter's story people are forced by the authorities on Earth and Mars to settle on a planet orbiting Proxima Centauri. They build shelters and harden their bodies to the radiation by taking vitamin supplements, injecting atropine and hoping for the best. The atmosphere of the planet, which is named 'Proxima c', christened Per Ardua in the book, protects the surface from some of the radiation, but when two of the characters are caught outside as a huge flare erupts from Proxima Centauri, just 6 million kilometres from Per Ardua, they have to urgently hide by climbing inside stromatolites that the native lifeforms also use to shield themselves from the radiation.

## A matter of time

Despite the flares, Baxter's Per Ardua is habitable. His characters explore all across the dayside, with Proxima Centauri rising higher in the sky as they near the sub-stellar point.

Supposing that it were possible to live on a tidally locked world like Per Ardua, we would not find a world as comfortable as Earth. Sure, it might have land and oceans, a breathable atmosphere and clement temperatures at least in certain parts of the world, but it would all feel quite exotic, quite alien, on a planet unlike our own. On Earth, the cycle of day and night is baked into our psyche and our biology. On an unchanging world without sunrise and sunset, how would we be able to measure time? And how would our body know when it is time to sleep and time to wake? On a tidally locked world, we'd be lost at sea, temporally, which is exactly the genius insight that Charlie Jane Anders had when writing *The City in the Middle of the Night*.

'I don't think it is healthy or possible for us to live without sleep cycles,' says Anders: 'I really just started obsessing about once you confront the idea of maintaining sleep schedules, how far are you willing to go into social control?'

Her novel's protagonist, Sophie, lives in the city of Xiosphant on a tidally locked planet called January. 'Everything in Xiosphant is designed to make us aware of the passage of time, from the calendars, to the rising and falling of the shutters, to the bells that ring all over town,' writes Anders in her book. 'Everyone always talks about Timefulness.'

Bells ring out to warn that a sleep period and accompanying curfew is about to begin, then the shutters automatically slam shut, and inhabitants are not permitted to do anything but sleep in this period. Being caught outside by the patrols of the authoritarian police force results in strict penalties and even expulsion from the city. Another city, called Argelo, imposes no restrictions whatsoever: the inhabitants are able to do whatever they want, and sleep whenever they want, 'But that leads to a weird chaos and a city run by gangsters,' says Anders.

The book is a favourite of Adiv Paradise, the planetary scientist from the University of Toronto whom we first met in the previous chapter; he highlights the way that Anders's novel describes how the environment of the planet impacts upon the societies that have settled on it: 'It's a fantastically imaginative novel,' he says, 'What I liked about it is that not only is Anders grappling with the implications of a planet like this, she's really using it as a setting to explore cultural questions.'

When discussing exoplanets, societal concerns are rarely top of the list. We're told whether it would be possible for life to arise and flourish on an exoplanet, but not how the exoplanet itself might affect the cultural development of any inhabitants, either native or non-native. Maybe this is the role of SF. In Anders's novel, the type of planet that January is has a dramatic effect on societal and cultural norms, changing the way that people live.

Paul McAuley describes his experience of visiting the Arctic Circle, and even living in northern Scotland, as giving some idea of what it would be like to live on a red-dwarf planet with permanent day or night, or at least a slow transition between the two, as is the case on Mangala.

'In Sweden, above the Arctic Circle, the locals were buzzing at all hours because it was light again – they get all their light at once,' he says. 'It would be a weird experience to have light all the time, or to live in perpetual sunset on the twilight strip, and you have darkness on your back and light at your front, and the shadows always falling one way and they don't move.'

In *The City in the Middle of the Night*, Sophie even takes advantage of these unmoving shadows, memorizing where they are so that she can hide in them to evade the security patrols.

While on January the shadows don't move, Isaac Asimov did find a way for the shadows to change on a tidally locked world. In *Lucky Starr and the Big Sun of Mercury*, the tidally locked Mercury librates. Think of libration as an apparent 'rocking' of a planet or moon on its axis as it follows its elliptical orbit around its parent body, which in Asimov's story results in a periodic half-mile movement of the terminator – fascinating! This means that the inhabitants of a base located near the changing terminator move in and out of night, creating a kind of day–night cycle that January so desperately lacks.

Libration is the consequence of an elliptical orbit. All planets orbit their star in an elliptical – rather than perfectly circular – orbit, something we've known since Johannes Kepler developed his laws of planetary motion in the early seventeenth century. In an elliptical orbit a planet's path around its star is slightly wider in one direction than in the other, a bit like an egg. Usually, the orbits of the planets are only slightly elliptical – Earth's orbit brings us as close as 147.1 million kilometres and as far as 152.1 million kilometres to the Sun. Mercury's orbit is the most eccentric of all the planets in our solar system. The Moon's orbit around Earth is also eccentric, and it too is seen to librate. In fact, it librates so much that we can see more than half the Moon from Earth at different times, amounting to a total of 59 per cent of the lunar surface – not bad for an object that supposedly always shows the same face to us.

To understand why libration occurs, think of the Moon's elliptical orbit around Earth – sometimes it is lagging a bit behind us, so we see around to part of the far side in one hemisphere, and other times it's a bit ahead of us, so we can see around the limb on the other hemisphere. Mercury also librates, as confirmed by NASA's MESSENGER (MErcury Surface, Space ENvironment, GEochemistry and Ranging) mission that operated in orbit around Mercury between 2011 and 2015, so Asimov got that right sixty years earlier. However, if a tidally locked world orbits on a near circular path, then it won't librate much. That is evidently the case for January.

Speaking of eccentric orbits, tidally locked hot Jupiters often have an orbit considerably more eccentric than even Mercury's. This results in things livening up a bit as they rush in close to their star before moving away again. The archetypal hot Jupiter for this is HD 80606b, which orbits a Sun-like star 190 light years away. As it gets close to its star, the temperature rapidly rises, generating global storms that wreak havoc within HD 80606b's atmosphere. NASA's thermal-infrared-seeking Spitzer Space Telescope has actually seen this in action, after a fashion, detecting HD 80606b heating up and triggering a fierce storm that ripples outwards from its sub-stellar point as it gets close to its star.

Could the intense heating of red-dwarf planets at sub-stellar points also generate storms as air rushes from the high-pressure dayside to the low-pressure night side? 'The instinct is to think that,' says Wordsworth, 'But actually the opposite could be true.' This is because tidally locked planets, with a length of day equal to their length of year, rotate slowly enough that their global atmospheric circulation is more gentle. After the relative rush of warm air to the nightside, 'You would expect the night air to be quite stagnant, circulation-wise,' Wordsworth notes. According to atmospheric models, an inversion layer would form, meaning that rather than temperatures decreasing with height, they would start to increase above the inversion layer.

'So the temperatures in the atmosphere over the night-side would be warmer higher up and colder towards the surface, the opposite of what we're used to,' says Wordsworth. 'The nightside would be this very eerie place where it is very still, and very, very cold and inhospitable at the surface.' That tracks with the depiction of the red-dwarf planet in *The City in the Middle of the Night*. Without giving too much away about the story, the eponymous city is an underground alien city on the permanent nightside of January, which has a frozen hemisphere that is deadly cold to humans.

During Anders's story, her protagonist Sophie has to escape from the city of Xiosphant and journey with her best friend and a group of smugglers to the next city on the twilight strip, Argelo. On the way there they must cross a body of water, the menacingly named Sea of Murder – so called because of the treacherous conditions, sea monsters and pirates. Where the Sea of Murder meets the blazing sun on the dayside it evaporates in a dramatic wall of hissing water vapour, only to precipitate back out as sleet and snow just beyond the day–night terminator. As we now know, this would only happen on a ribbon world, and atmospheric circulation should prevent most of the dayside becoming too hot, but at the time Anders wrote her story, it fit with what we thought we knew about these planets.

Indeed, when researching her book Anders reached out to the planetary scientist Lindy Elkins-Tanton of Arizona State University to ask a whole bunch of questions about tidally locked planets. 'I had a lot of conversations with Lindy, who was incredibly helpful,' says Anders. 'We talked through how the Sea of Murder might work, and it was deliberately very weird and I sort of assumed that there was some kind of underground spring or some other source of water feeding it. But I should say that if I got anything wrong it's not her fault! All the decisions were mine in the end.'

The presence of surface water raises an interesting question about what impact that water has on a tidally locked planet's climate. On Earth the oceans have a terrific impact on our climate, absorbing and redistributing heat, facilitating the development of winds and providing water for precipitation elsewhere on the planet. It's no surprise, then, that the distribution of water versus land actually has quite a lot of importance for the climate of a tidally locked planet too. Research by Adiv Paradise and his colleagues at the University of Toronto, including lead researcher Evelyn Macdonald, found that the overall climate on tidally locked worlds is strongly dependent upon the amount and location of land versus water on the planet's surface. In simulations, planets with a significant amount of land on the dayside had average surface temperatures up to 20 degrees Celsius higher, and the more land there was the hotter the dayside grew and the drier the atmosphere became, because the more land there is, the less surface water there is available to evaporate into the atmosphere. So we'd expect tidally locked planets with large land fractions to be desert worlds.

Want the best of both worlds? The simulations found that achieving the highest global temperatures possible while maintaining a large amount of moisture in the atmosphere requires a large ocean at the sub-stellar point, since the increased water vapour evaporating from the ocean leads to more clouds and a greater ability to transport heat to the nightside.

## Timeless

There are two things that writers of space opera like to include in their stories. One is magnificently ancient aliens, beings that evolved billions of years before anyone else – the Xeelee in Stephen Baxter's novels, the Inhibitors in Alastair Reynolds's *Revelation Space* series, the Shadows and Vorlons in *Babylon 5*. The other thing is blowing up planets and stars – just ask George Lucas and his favourite toy, the Death Star. And even if your home planet doesn't get blown up, eventually nature will have its way: your star will inevitably grow hotter as it ages, gradually scorching your world. Earth has about a billion years left before the Sun renders it too hot for life to survive and the oceans boil away.

If life wants to survive into the deep future, to live as long as those fictional ancient aliens, eventually the day will come when it has to find a new home.

The bulk of this chapter has been centred around the unchanging nature of tidally locked worlds around red dwarfs. No way to mark the passage of time, no sunset or sunrise, no seasons, no change in the weather. Now though, as we look to the deep future, this timelessness could be the saviour of life in the universe. That's because red dwarfs are the trees of the cosmos. While all the other stars and their planets burn out and die, withering away like flowers in autumn, the red dwarfs will keep going. They are incredibly long lived, able to survive for perhaps a trillion years. Our Sun's entire lifetime of about 10 billion years will be a fading memory by comparison; the short but dramatic lifespans of the most massive stars that explode as supernovae within a few million years of their birth a tiny fragment of time relative to the aeons that red dwarfs will experience. Red dwarfs have the capacity to outlast every other star because they're like fuel-efficient cars, taking mere sips of the hydrogen gas that fuels the nuclear fusion reactions in their core, whereas larger stars have to guzzle up their nuclear fuel to generate the energy required to counteract gravity. When the Sun and all the other stars have long gone, when new star formation has ceased simply because all the raw material in the universe has been used up, the red dwarfs will still be there, and that's where life will want to be if it aims to survive that long.

By that time the universe will be a much duller place, quite literally. The bright lights of massive stars will have switched off, and the only light will be the dim and vaguely pinky-red hue coming from red dwarfs. Most of a red dwarf's light output is actually in infrared light.

On Earth, green chlorophyll absorbs light and converts it into energy through the process of photosynthesis (sunlight plus water plus carbon dioxide equals biochemical energy for plants to live on, and oxygen for us to breathe). While there are a few other photosynthesizing pigments on Earth, green is overwhelmingly dominant, focusing on absorbing blue light, because blue photons carry the highest energy.

On a planet orbiting a red dwarf, any hypothetical photosynthesizing cyanobacteria (the tiny microbes that do the biochemical work in plants) can't afford to be picky about what photons they absorb, since they will all typically be of lower energy because of the cool temperature of the star. Leaves, or their extraterrestrial analogue, would have to be optimized to gather up every scrap of light that they can without reflecting any of it. Consequently, these 'leaves' would most probably be black. It's even been suggested by Scottish astrobiologists John Raven of the University of Dundee and Ray Wolstencroft of the Royal Observatory, Edinburgh, that photosynthesis is feasible with near-infrared light, which would be useful on a planet orbiting a red dwarf.

As an aside, Jack O'Malley-James, who studied under Raven while at the University of St Andrews and who is now at Cornell University in New York, suggests that plant life on planets orbiting not one but two stars could react in more complex fashion to the two different light sources. Supposing that the binary system contains a red dwarf and a more Sun-like yellow star, O'Malley-James describes in his 2011 academic research paper 'Life and Light: Exotic Photosynthesis in Binary and Multiple Star Systems' how there could be two tiers of vegetation on these planets, one tier adapted with black leaves suited to the redder-hued light of the red dwarf and the other group of plants evolving to photosynthesize light from the yellower star, with green or even red leaves.

Double, or binary, star systems such as Tatooine in *Star Wars* are our topic of choice in the next chapter. Given their ubiquity, red dwarfs undoubtedly make up a significant proportion of binary members.

Despite their abundance, despite their unvarying nature, red dwarfs and their planets are absolutely, seductively fascinating from a scientific point of view, and ripe with storytelling potential for science fiction. Amaury Triaud hopes that SF will be able to convey that planets orbiting red dwarfs can be interesting, and potentially be habitable. And whereas Charlie Jane Anders had noted a dearth of fiction based around such worlds, that may now be changing in support of Triaud's hopes, perhaps even inspired by Anders's *The City in the Middle of the Night*. Emma Puranen, who as we have seen is studying trends and connections between types of exoplanet in science fiction, has noticed more tidally locked planets in SF of late. 'It's a concept that's very much in vogue right now,' she says.

It might just be time for red dwarfs to shine.

# 7

# LANDS OF THE RISING SUNS

At first glance, Luke Skywalker's homeworld of Tatooine doesn't appear too different to Arrakis from Frank Herbert's *Dune*. They are both desert worlds, nary a drop of water to be seen on either planet. In describing Tatooine, Luke explains that 'if there's a bright centre to the universe, you're on the planet that it's farthest from,' a somewhat clumsy way of saying he thought Tatooine was pretty dull. Slug-like mafiosos, Jedi hermits and hives of scum and villainy aside, maybe he had a point, except for when each day ended. Then the true beauty of Tatooine was revealed in the hues of a double sunset of a binary star.

Earth orbits a single star, our Sun, but many stars in the universe are actually found as doubles and triples, or even occasionally quadruples, quintuplets and sextuplets. Within 32.6 light years (10 parsecs – a measure of distance, not time, as *Star Wars* infamously got wrong) of Earth there are 317 star systems, and of these, 85 systems – 27 per cent – have more than one stellar component. Overall, in the galaxy, about two-thirds of stars are found in binary systems.

Often, a planet in a multiple star system will orbit just one of the stars, with the other star or stars at a much wider distance. The planet wouldn't really have two suns in the sky, since the second star would be far enough away that it would appear as no more than a very bright star. For instance, the Sun appears thirty times smaller in Neptune's sky as it does in Earth's sky. A second star (at least the size and temperature of our Sun) at the distance of Neptune therefore wouldn't have a great impact on Earth. Similarly, in many multiple star systems, the other star or stars are too far away to really affect the climate or give double sunsets on a planet orbiting just one of the stars.

Things change in the situation of a planet orbiting two stars, like Tatooine does. The two stars are so close together, and so close to the planet, that they appear as two suns in the sky. Indeed, 'Tatooine planet' is one of those rare phrases from science fiction that has actually entered the scientific lexicon to describe this scenario. There are many scientific papers describing potential Tatooine planets, and it's pretty cool to see how so many astronomers have embraced the reference to *Star Wars*. Yet we need to temper our expectations because so far Tatooine planets have been found to be elusive at best, and at worst, quite rare.

It's a puzzle, no doubt about it. There are certainly enough double stars in the universe to make what we call 'circumbinary' planets – that is, planets that orbit two stars with romantic double sunsets, like Tatooine – fairly commonplace. Yet of the 5,700-plus known exoplanets at the time of writing, only about a dozen true circumbinary planets are known, and they are all gas giants like Jupiter rather than potentially habitable Earth-like planets.

'We've found no evidence so far that rocky planets like Tatooine exist,' says Amaury Triaud, the astronomer from the University of Birmingham whom we first met in Chapter Six. He pauses, then adds: 'But of course, that doesn't mean that they don't exist.'

## The suns of Tatooine

In theory, we know that a planet like Tatooine could exist, but whether such a world is real or not remains an unanswered question. Given what we do know about the possibilities for such worlds, how realistic is Tatooine's portrayal in the *Star Wars* universe?

On screen, *Star Wars* doesn't give us too many details, but in the 1977 novelization, credited to George Lucas but ghostwritten by Alan Dean Foster, Tatooine's suns are described as being G1 and G2 stars. 'G' refers to their spectral classification (remember the mnemonic 'Oh Be A Fine Girl/Guy, Kiss Me', which we introduced in the previous chapter), describing their temperature and hence their colour. O-class stars are the hottest stars, M-class stars are the coolest and G-class yellow stars are roughly in the middle, with a surface temperature of about 5,500 degrees Celsius. The numbers after the classification – 1, 2 and so on – refer to slight graduations in the scale; one of Tatooine's stars is slightly more massive and hotter than the other. Earth's Sun is a G2 star, so Tatooine's suns are both very much like our own.

Foster evidently did his homework; he refers to the two stars orbiting a common centre of mass between them, and Tatooine orbiting around them far enough out to allow a stable, if hot, climate. Essentially, he's describing Tatooine as being near the inner edge of the circumbinary habitable zone. It's worth pointing out that Tatooine isn't hot and dry because it has two suns – there's always going to be a distance from those stars where temperatures are suitable for liquid water and habitability, it just depends upon how far from those stars your planet orbits, same as with planets around single stars.

The description in the novelization matches what we see on screen in *Star Wars*, where a wistful Luke looks out at the sunset, as we see two suns of roughly equal size, one appearing a little bloodier than the other as a result of atmospheric reddening lower down in the sky. (As an aside, this is why we get red sunsets on Earth. Near the horizon sunlight has to pass though more of the dusty atmosphere. The atmosphere preferentially scatters blue light, hence why the sky usually appears blue, but nearer the horizon sunlight has to pass through a thicker column of atmosphere to reach our eyes: all the blue light is therefore scattered away, while the red light passes through relatively unhindered.)

'One of my former students played around a bit to see if Tatooine was accurate, and if you assume that the two stars are like our Sun, then Tatooine would be in the habitable zone and the sizes of the stars in the sky are approximately correct,' says Triaud.

Whether or not a planet has a stable, habitable climate depends in large part on the size of the planet's orbit. Too short and it will burn in the close proximity to the heat of its star(s); too wide and the planet will freeze. If it's at just the right distance, however, then the planet's temperature will fall within a range suitable for liquid water to exist on its surface. Earth is at just the right distance from our Sun for liquid water to exist on its surface in great quantities, but habitability doesn't always mean there's lots of water. Like *Dune*'s Arrakis, Tatooine seems to fall into that category. Earth only has one Sun, so things are pretty straightforward, but how would having two suns complicate matters of habitability?

'If the stars are close enough to each other, then it's like having one heat source,' says Knicole Colón, who if we remember is an astrophysicist at NASA's Goddard Space Flight Center and the deputy project scientist for the TESS mission. The tightest binary stars can have separations of just a few million kilometres, orbiting each other in a matter of days, and from a circumbinary planet's point of view, the two stars might as well be a single star, since any changes in heat as one star moves in front of the other will be minor and smeared out across a seasonal average. If we take the double sunset that Luke watches as typical for Tatooine, then it implies that its two suns are on fairly tight orbits, leading to a stable climate for the desert planet.

If the binary stars are more widely separated, all sorts of shenanigans can lead to climate havoc. The interplay between the physical separation of the stars and the orbit of the circumbinary planet will lead, at times, to the planet being much closer to one star than the other. But the consequences of this rather depend upon the individual masses of the stars.

'The closer in mass the two stars are, the more likely you are to have a planet that is dipping in and out of the habitable zone, leading to interesting effects and temperature changes,' says astronomer Grant Kennedy of the University of Warwick.

Let's deconstruct this. The stars are orbiting a common centre of mass, and the planet is orbiting around them. At certain times, the two stars will be adjacent to one another relative to the planet; they'll be at the same distance from the planet, but at their widest separation on the planet's sky. Nonetheless, the amount of heat the planet receives from each of its suns will be even when they are both aligned in this fashion.

At other times, however, as the two stars orbit their shared centre of mass, one of the stars will be significantly closer to the planet than the other in real terms, while at the same time the two stars will be at their closest apparent separation in the planet's sky (the two stars and the planet will almost be in a line). Depending upon how the orbital plane of the planet is aligned with the orbital plane of the stars, the stars could even eclipse and occult one another, or if not, end up appearing just above or below each other on the sky.

Now the mass, and therefore the temperature, of the stars becomes the crucial factor. If the stars are significantly different in mass – say one is a G-type star like our Sun, or Tatooine's two suns, and the other is a cool, red M-dwarf – then the hotter, more massive star will always play the dominant role. In this example, that would be the G-type star, so it wouldn't matter too much if it was the closest or furthest star to the planet, since the red dwarf's heat would be too feeble to make an impact. Of course, when the G-type star is furthest from the planet, the planet would receive less heat, resulting in a winter season that could potentially be quite severe.

However, if the stars are similar in mass and sufficiently widely separated, this could mean big trouble for the planet because each star will contribute a significant amount of its own heat. We can envisage several scenarios. When the planet is at the same distance from both of its stars, it will receive equal amounts of heat from them, which will affect the planet's climate to different extents depending on how far the planet is from the stars. Too close to both stars, and the planet might be too warm, or maybe it's just at the right distance to fall into the habitable zone. Then the planet moves around the stars, and the stars move around each other. If the planet is too warm when equidistant to both stars, then when it is closer to one star than the other, it may then be cool enough for a period of habitability on the surface. Or, if the planet is habitable when equidistant from the two stars, it may become too cold when close to only one star. The point is, although the planet is orbiting the two stars on a fairly circular orbit, the planet is not always in the habitable zone. Depending upon the subtleties of the circumbinary planet's orbit, it could lead to summers that are too hot for life to be comfortable, and frequent winters that are too cold, resulting in interruptions in the habitability of the planet.

'We've seen systems where the planet could be in the habitable zone some of the time, so what does that mean?' asks Colón rhetorically: 'Could there be drastic seasonal changes? Maybe the planet is habitable part of the year, and the rest of the time it's in a frozen ice age, and then how can life deal with that, could it live underground? There are all these possibilities, and lots of climatic havoc could be wreaked! But the level of havoc depends upon a lot of factors.'

We do know that there are seasonal changes on Tatooine. Luke's uncle, Owen Lars, talks about seasons and harvests. The Lars family own a moisture farm, which they live on, harvesting the small amounts of water vapour in the atmosphere. Rain is a rare and notable event, happening maybe once per year or once every few years. When it does rain, small crops can sprout from the sand, as Foster describes in the novelization. The fact that there are seasons and rare rainfalls implies that there are periods when the relative humidity increases. Although it is feasible that these seasons could be partially linked to the tilt of Tatooine's axis of rotation, similar to the cause of the seasons on Earth, they could also be connected to warmer and cooler periods depending upon which of its suns the planet is closest to at any given time.

A fictional planet in a binary star system that definitely does have seasons is Helliconia from Brian Aldiss's trilogy of novels *Helliconia Spring* (1982), *Helliconia Summer* (1983) and *Helliconia Winter* (1985). The planet of Helliconia isn't quite circumbinary; it orbits a fictional G-type star called Batalix every 480 days (where a day is nearly 26 hours long), and Batalix in turn orbits another star, the hot A-type star Freyr, every 1,825 Helliconian years (a 'great year', equating to about 2,500 Earth years). Because A-type stars are bright and hot, Freyr is the dominant source of energy in the system; Helliconia's distance from Batalix means that without the extra heat from Freyr, Helliconia's temperatures barely rise above freezing. You can probably see where this is going: Helliconia and Batalix's long, elliptical sojourn around Freyr results in exceptionally long seasons that are centuries in extent. The seasonal changes can wreak havoc on the native civilizations (there are two, the phagors and an alien species that looks and acts remarkably human, like many of the aliens in *Star Trek*) and play a vital role in the trilogy's story, which spans the majority of a 'great year' as we watch the two civilizations rise and fall.

Helliconia is a favourite of SF author Stephen Baxter. 'It's written beautifully, with sweeping seasons and civilizations washing backwards and forwards, it's fantastic,' he says. 'That's because it's so well worked out, and Aldiss had help from the late Jack Cohen (a reproductive biologist) and Ian Stewart (a mathematician).'

Besides their academic backgrounds, Cohen and Stewart were both SF authors in their own right (they also co-wrote the very good book *Evolving the Alien: The Science of Extraterrestrial Life* (2002), which does for aliens what this book hopefully does for planets, even if they are snooty towards televisual and cinematic SF). Aldiss consulted with them throughout the writing of the trilogy as they sought to add a plausible scientific veneer to Aldiss's epic. 'They did a great job with Aldiss, who kept coming up with these mad visions, which they then rationalized,' says Baxter. 'It all worked very well.'

## The Three-Body Problem

There are other effects that twin suns might have on a planet's short-term climate. Light levels, and the colour of the sunlight, might change depending upon which of the two suns is most prominent in the sky at the time, which could affect things such as plant growth.

The number of hours of daylight would also change depending upon how widely separated the stars are. The two stars will never get too far from one another in the circumbinary planet's sky, roughly 20 degrees (about forty times the diameter of the full Moon in our sky) at most, says Triaud. Some days will be longer, when the stars are at maximum separation, and others will be shorter, when the stars are closer in the sky. This wandering of their relative positions will also affect the timings of sunrise and sunset.

Some planets are found in systems with not just two stars, but three. One classic example in fiction that we must mention is Cixin Liu's popular novel *The Three-Body Problem* (2008) (and its sequels, *The Dark Forest* (2008) and *Death's End* (2010), plus two television adaptations, specifically a Chinese production (2023) and the Hollywood version from the makers of *Game of Thrones*, as shown on Netflix (2024–)). The novel's title directly refers to a planet in a triple-star system, and how that planet's proximity to its three stars at different points in its orbit plays havoc with the climate and the life that calls that world home. Although it is not explicitly stated which star system it is, there are enough clues to suggest that the author was thinking of the alpha Centauri system, which is the closest star system to our Sun. However, the fictional orbital mechanics don't match the reality of the alpha Centauri system, which features two stars – G-type alpha Centauri A and the slightly cooler K-type alpha Centauri B – that orbit around one another at a distance of 4.3 light years from the Sun. Then there's a third star that should now be familiar to you, as it has featured in previous chapters: Proxima Centauri, a cool, red M-dwarf that orbits the main pair on a very long orbit of about half a million years, and which is actually the closest star to the Sun, at a current distance of 4.2 light years. Furthermore, as we know, there are planets in this system! They are not circumbinary, however – the three that have been discovered so far orbit only Proxima, including Proxima b in the habitable zone. No planets are yet known around alpha Centauri A or B, although, at the time of writing, the possible signature of a planet orbiting alpha Centauri A has been tentatively identified and is being followed up on by NASA's James Webb Space Telescope, while in 2012, it was thought that astronomers had discovered a planet orbiting alpha Centauri B, only for this to turn out to be an error caused by spurious noise in the data. Alpha Centauri A and B have a quite eccentric, or elongated, orbit around one another; they get as close as 1.676 billion kilometres (a little more than the distance between Saturn and the Sun) and as far from one another as 5.326 billion kilometres (about the average distance between Pluto and the Sun), so they are too widely separated for circumbinary planets to form.

Astronomers have actually found strong evidence of a planet that orbits three stars, however. GW Orionis is a young star system 1,300 light years away, near the Orion Nebula, which is a large star-forming region visible to the naked eye as a faint smudge beneath the three stars that form the Belt of Orion in the winter constellation of the Hunter. A team of astronomers led by Stefan Kraus of the University of Exeter used the Atacama Large Millimeter/submillimeter Array (ALMA), a group of 66 radio dishes in the Atacama Desert in Chile, to observe three dusty rings encircling the triple stars of GW Orionis at distances of 6.7 billion, 28 billion and 51 billion kilometres, respectively. The inner ring contains enough material to build thirty planets, each with the mass of Earth, and a warp in the plane of the rings has been attributed to an unseen giant planet. Although there is still time for some planetary migration to take place, the scale of this planetary system is far larger than our own.

As mentioned earlier, stars don't just come as singles, doubles and triples; occasionally there are also stellar quartets, quadruplets and sextets. Could planets exist around this many stars? Certainly, in fiction, this has been broached. The basic plot of Vin Diesel's film *Pitch Black* (2000; dir. David Twohy) sees trouble arise when darkness falls on a planet with multiple suns and carnivorous monsters come out to play. This idea of rare spells of darkness on planets with multiple suns riffs on a much older tale, specifically Isaac Asimov's short story 'Nightfall' of 1941 (adapted into a novel by Robert Silverberg in 1990), which is considered to be one of the greatest science-fiction short stories ever written. The story is about the inhabitants of a planet called Lagash that orbits six stars, and the relative orbits of the stars mean that Lagash is in constant daylight, with at least one of its suns always up in the sky. Consequently, its inhabitants have never experienced the night or seen the stars beyond the half-a-dozen suns that Lagash orbits. Then, one day, Lagash's moon, which cannot be seen in the glare of the daylight and is not known to Lagash's population, eclipses one of the six suns when that sun happens to be the only one in the sky at that time, and for the first time in two millennia, the sky grows dark and all the stars come out. Faced with the true immensity of the universe beyond their own system, the inhabitants of Lagash are driven mad, lighting fires and burning their cities to the ground in a crazed effort to try and bring light back to their world.

No planets have been discovered in a sextuplet system, but there is one remarkable five-star system that features not one, but two planet-forming discs. Located 450 light years away, the system is called GG Tau, in the constellation of Taurus, the Bull. Three of its stars form what is known as a hierarchical triple – two of them make a binary system orbited by a third star. One of the planet-forming discs surrounds the brightest star in this binary, while the other disc encircles all three stars. Then beyond the disc are the two remaining stars in the system, forming their own binary. Being only 1.5 million years old, the discs haven't had time to form fully fledged planets yet, but several hotspots in the discs point to the presence of protoplanets.

'GG Tau is an unusual system,' says Kennedy: 'It highlights that discs are not too fussy about where they exist, be it in circumstellar, circumbinary or circumtriple orbits.'

Technically, any planets that do form in the discs won't have five suns, but it could lead to the very odd scenario of some planets orbiting three stars and others, closer in, orbiting just one. What is notable is that detecting dusty and potentially planet-forming rings around young stars such as GG Tau can sometimes be somewhat easier than spotting individual evolved planets around older stars. Dusty discs stand out at longer wavelengths, and astronomers have grown adept at using ALMA to discover them. Indeed, ALMA's observations have led to a discovery more bizarre than any circumbinary planet in science fiction.

## Looping planets

In 2019 Kennedy's own research led to the finding by ALMA that the discs around young, wide binaries are sometimes tilted with respect to the plane on which the two stars orbit each other. (Wide in this context means that the two stars take at least one month to complete one of those mutual orbits.) The fact that the dusty discs are tilted – often by as much as 90 degrees, so the disc ventures over the poles of the stars – means that any planets that form in the disc in the future will also orbit over the poles, rather than around the stars' equatorial planes (which is the ecliptic in our solar system).

While it is difficult for astronomers to go back and trace what caused the discs to become tilted, computer simulations can paint a picture of what might have happened.

'There might have been a binary system that formed with a disc around it, and then another star system might have come in and had a reasonably close encounter, which could have imparted a torque on the disc and rotated it,' suggests Kennedy. Much tighter binary systems would be able to resist such perturbations because of the mutual gravitational attraction between the stars, which keeps them and their surrounding planet-forming disc of gas and dust in line.

'The weird thing about polar alignment is that it's more stable the more eccentric [that is, the further it is from being circular] the binary's orbit is, so you don't always need to give the disc too much of a perturbation,' adds Kennedy: 'As a result, if you have a wide and eccentric binary with a disc around it, then it's just as likely that the disc will be perpendicular as it is coplanar.'

Based on what we see on-screen, Tatooine's two stars seem fairly close in the sky and probably do not form a wide binary, so we would expect Tatooine to orbit in their equatorial plane rather than in a polar orbit. Suppose, though, that Tatooine's stars were a wide binary. Given what we know from Kennedy's research into wide, young binary systems in our own Milky Way galaxy, there would be a decent chance that in this case, Tatooine would be on a polar orbit. How would this affect the inhabitants of Tatooine?

'Imagine that we were on Tatooine,' suggests Kennedy: 'In a coplanar system we would see its suns moving back and forth in the plane of the ecliptic, whereas in a polar system the suns would be seen to move up and down relative to the ecliptic, but they would also still be coming towards and away from us over the period of their orbit.'

So, in practical terms, there would still be double sunsets and sunrises, although they would perhaps be less frequent. In addition, however, there could also be something else in the sky with a beauty to rival that of a binary sunset: incredible aurorae borealis.

If you have never seen the northern lights before, then make sure to include them on your bucket list and, at some point, venture up to the Arctic Circle to witness them in all their splendiferous glory. No magical phenomenon, they are the result of an interaction between our planet's magnetic field and the solar wind of charged particles emitted by our Sun. The charged particles – electrons, protons and charged atoms, or ions – spiral around magnetic-field lines that emanate from Earth's magnetic poles. They follow the magnetic-field lines down into the atmosphere above the poles, where they collide with atoms of nitrogen and oxygen with such energy that they cause the gases to glow, and we see the wonderful dance of the northern (or southern) lights in the night sky.

Although the solar wind is ever present, particularly fast gusts can escape from our Sun through a phenomenon called coronal holes. On the visible surface, or 'photosphere', of the Sun, magnetic-field lines emerge from active regions (which are often home to sunspots) and loop up, with both ends anchored in the photosphere's plasma. These are closed magnetic-field lines, but in coronal holes the magnetic field lines are open, one end protruding into space. The fast solar wind is able to blow through these holes, which can come and go on a regular basis. Most importantly for this discussion, large coronal holes are more commonly found near the Sun's poles.

Now we see how this could impact circumbinary planets if they're in a polar orbit. As they pass over the poles of their suns, they would be in full view of the coronal holes and receive the full blast of the fast stellar winds, and this would lead to dramatic aurorae. If at a safe distance, then the circumbinary planet's inhabitants would just enjoy a good light show, but worlds that orbit closer to their stars, such as tidally locked planets, will feel the brunt of the fast wind. If the planet lacks magnetic defences, the wind could blow the planet's atmosphere completely away into space, resulting in a tail of gases like a comet's tail streaming behind the planet.

## Where are all the Tatooines?

The discussion of what might happen to planets on polar orbits too close to a star raises a good question. Just how close can a circumbinary planet get to its binary stars?

'Binary stars have a region of instability around them that means you cannot have a planet closer than four times the binary period,' says Triaud. In other words, if two stars orbit around one another with a period of fifteen days, then no circumbinary planet around them can have an orbital period (a 'year' on that planet) shorter than sixty days. If they were any closer, with a shorter orbital period, they would be slingshot out of the system by their two stars. 'Most of the circumbinary planets that we've found so far are six times the binary period,' says Triaud.

However, the statistics could be misleading, built up as they are from a relatively small sample of circumbinary planets. We don't yet know whether the scarcity of circumbinary planets is because they are genuinely rare, or because finding them is difficult, thanks to the second star throwing up so many complications that hinder the detection of the planets. Often, if an astronomer hunting for exoplanets sees a close binary star system, they steer well clear, not wanting to give themselves that particular headache.

'There is a bias against finding planets in binary systems,' admits Kennedy, but not everyone is put off: 'Amaury Triaud is really interested in circumbinary planets. He has a radial-velocity programme, and I think that sort of work is really important.'

At the time of writing, Triaud and his colleagues had several science papers in the works detailing new circumbinary planet discoveries, including one detected by the transit method from a telescope in Antarctica. 'I don't know if you can say that the floodgates have opened, but it has become possible to detect them,' says Triaud.

All the circumbinary planets discovered thus far have been giant planets like Jupiter. Examples include Kepler-16b, which was discovered orbiting two small stars 200 light years away from Earth by a team led by Triaud, and it has a mass similar to that of Saturn. With an orbital period of 229 days, placing it beyond the outer edge of the habitable zone, Kepler-16b is one of the widest orbits found so far for a circumbinary planet.

Another Kepler Space Telescope discovery is Kepler-34b, which is a gaseous world 6,200 light years away. It has one-fifth of the mass of Jupiter and is seen to transit both of its Sun-like stars. The Kepler-47 system is notable because it has three circumbinary planets, all low-density gas bags. One of the planets, Kepler-47c, is in the habitable zone of its two stars, but without land or ocean it is unlikely to have life – at least, as we know it. Meanwhile, Kepler's successor, the Transiting Exoplanet Survey Satellite (TESS), has identified the circumbinary planet TOI 1338b about 1,320 light years from Earth. Like all the other known circumbinary planets, TOI 1338b is also largely gaseous, analogous to Saturn in size.

Where are all the small, rocky circumbinary planets like Tatooine? Maybe there's something about binary systems that prevents their formation. Or 'it could be that there's a lot of small planets that we're not seeing,' says Knicole Colón. This would be simply because small exoplanets are much harder to detect in general.

The region of instability around a binary star may have some say in this. Based on what we see in our solar system, and what detailed computer models of planet formation tell us, we expect small, rocky planets to typically form close to their star, since that is where gravity draws the heavy elements – silicates, iron, nickel – required to build such planets. In many cases the region of instability might coincide with the area in which terrestrial worlds would typically form, and therefore the zone of instability would clear out any dusty discs before planets could form.

Another way to make a larger solid planet might be through evaporation: as we alluded to earlier, and in previous chapters, astronomers have discovered many cases where a giant planet has migrated so close to its star that the star's radiation has heated the planet's atmosphere to the point that the stellar wind can strip the atmosphere away, evaporating it. Remove the atmosphere and all that's left behind is the large rocky core of a planet. Around binary stars, however, the zone of instability would prevent planets from migrating too close to their stars to evaporate – if they got too close, they'd receive a gravitational slingshot that would fling them out of the system and into interstellar space.

This actually leads to several predictions. One relates to a puzzling discrepancy that astronomers have noticed, which is that there is a dearth of exoplanets with radii between 1.8 and 2.4 times the radius of Earth (which is 6,371 kilometres). We considered one explanation for this in Chapter Four, which is that the evaporation of planets by the stellar wind can allow them to jump across this gap, transforming from a large gas-rich world more than 2.4 times the size of Earth to a smaller and comparatively airless world with less than 1.8 times the radius of Earth. 'But that's less likely to happen for circumbinary planets,' says Kennedy, so the prediction is that binary systems should have more worlds larger than 2.4 times the size of Earth than worlds smaller than 1.8 Earth radii because they wouldn't have the chance to jump across that gap. Indeed, this is what results so far have shown.

The other prediction is Amaury Triaud's 'suspicion that we will find mostly water worlds around binary stars'. This is because, as we have said, the zone of instability would not permit rocky planets to form close in, but icy worlds that form further out, beyond the habitable zone, could migrate inwards and thaw as they get closer to their twin suns, just as we saw in Chapter Four.

This would seemingly suggest that a planet like Tatooine will remain a world of fiction, although despite his prediction, Triaud is hesitant to completely rule it out. 'I don't want to exclude the presence of rocky planets, since nature is always surprising.'

## Planet Gemma

Although Tatooine steals all the headlines, Triaud's favourite fictional circumbinary planet is one he helped create – Gemma (pronounced with a hard 'G') from Laurence Suhner's *QuanTika* trilogy of novels. Currently, no publisher has taken a chance on translating the books from their original French, although Suhner has made a self-published English version of the first book in the trilogy, named *Vestiges* (2012), available online.

The novels are based around the discovery of the remnants of an ancient civilization that once visited Gemma. Suhner was keen to include accurate science in her writing and went to great lengths to ensure it. Already having an archaeological background, Suhner actually took a physics degree at the University of Geneva so that she could write and understand the science in her novels better. When it came to the binary stars around which Gemma orbits, Suhner turned to astronomers at Geneva Observatory, where Triaud was working at the time, and asked for help in describing the system. The Geneva astronomers built a model of Gemma and its binary-star system that was scientifically accurate.

'Funnily enough, I wasn't working on circumbinary planets at the time, and now I am,' Triaud ponders with amusement. 'In so many strange ways, science fiction becomes reality!'

Perhaps science fiction stole a march on science research when it came to circumbinary planets because the idea of having two suns – or more – seems so alien to us. Life on Earth evolved under the gentle radiance of one sun, adapting to the regular diurnal cycles and the predictable procession of the seasons. When something is so far outside our experience that most people would not even think that it could be any other way, that's usually what provides fertile grounds for science fiction. And yet, for a time, some scientists thought that even our Sun could be a binary star.

Of course, Earth and the other planets are not circumbinary. But what if there were a second star that orbited at great distance around the Sun? If it were a faint M-dwarf, perhaps located 1 light year away, it might not be obvious at all. Instead, the purported evidence for a binary companion to our Sun came from controversial suggestions that there was a periodicity to the mass extinctions on Earth. This led several scientists, including Daniel Whitmire of the University of Louisiana at Lafayette and Al Jackson of the Lunar and Planetary Institute in Arizona, to propose that there could be a stellar companion to the Sun leisurely orbiting around the edge of the Oort Cloud of comets far beyond Pluto, its gravity disturbing those comets and periodically sending them towards the inner solar system where they in theory collide with the planets, including Earth, causing extinctions. Because it was seen to bring death from above, the hypothetical star was nicknamed Nemesis.

It's an intriguing idea, but alas, the evidence hasn't held up. The periodicity in mass extinctions has been contested, partly because palaeontology is not always an exact science when it comes to dating things that have been buried in the ground for hundreds of millions of years. More damning is that despite careful searches, astronomers have not found any red dwarf star closer than Proxima Centauri. The observations of NASA's Wide-field Infrared Survey Explorer, or WISE, were the final nail in the coffin for the idea – WISE scanned the entire sky and would have found anything larger than the planet Saturn lurking in the furthest reaches of the solar system, but it found nothing. (There could still be planets smaller than Saturn out there; attention has now turned towards searching for a hypothetical planet referred to as 'Planet Nine', which has been contentiously invoked to explain the puzzling orbits of some minor bodies in the outer solar system.)

The threat of mass extinction every 27 million years or so notwithstanding, how exciting would it be to have a second star in the solar system? 'This is where it is fun from a science-fiction perspective,' enthuses Triaud: 'If there were a second star in our solar system, then it would mean we would have a shorter travel time to the nearest star, possibly with an entirely different set of planets, geology and life.'

Of course, this would only apply if the planets weren't circumbinary, and if the individual stars in a binary system had their own family of planets around them. There's still the nagging feeling that circumbinary planets, or at least rocky ones like Tatooine, are going to be rare. We'll soon know the answer, thanks to projects such as Amaury Triaud's search for circumbinary planets, expensive space missions such as NASA's TESS and future missions such as the European Space Agency's PLATO space telescope, set to launch in 2026, which by working in combination will discover thousands of exoplanets.

'TESS is now building up the statistics, but there's still only tens of these circumbinary planets out of more than 5,000 exoplanets discovered in total,' says Colón in her summing up. 'So they are out there, but they're relatively rare and harder to find, though maybe not so rare in the sense that it's not crazy for George Lucas to have come up with one like Tatooine.'

# 8

# EXOMOONS

'Wow, is that pretty.'

It was Christmas Eve 1968, and three men were the furthest humans from home. Jim Lovell, William Anders and Frank Borman were the crew of Apollo 8, the first crewed voyage to the Moon. They didn't land, but rather circled our nearest neighbour for twenty hours. On their fourth orbit they emerged from around the lunar far side to be greeted by a view of Earth, half draped in shadow, rising above the Moon's craggy surface. All of humanity, barring those three brave men in their spacecraft, were 386,000 kilometres away. It was a profound moment, prompting Anders to reach for the camera and utter the words above. The resulting photograph, named *Earthrise*, is one of the most famous photographs in history. It's a perspective of Earth that less than three dozen people have ever had, but as NASA's Artemis lunar programme kicks into high gear, it's a view of Earth that hopefully more people will have in the future – a view that more people need to have. We perhaps don't always realize it here on the surface, but our planet is small and vulnerable amidst the expanse of space.

But not alone.

Our companion until the end of the Earth is the Moon. We're fortunate to have her, and that's not meant philosophically, but literally. The Moon is the result of a chance impact, a collision between a small protoplanet – kind of an embryonic planet that hasn't fully grown yet – and our young Earth. Scientists have even given the doomed protoplanet a name: Theia. Around 4.4–4.5 billion years ago, when Earth was no older than 100 million years, the Mars-sized Theia wandered into our sphere of influence and struck Earth a glancing blow. Theia was obliterated, while the entirety of Earth's crust and much of our mantle were ripped off by the titanic impact. Quickly, the debris settled into a molten ring around our bruised world and coalesced into our Moon. But to do so the impact required fairly specific conditions to avoid sending all that debris out into deep space where it could not form the Moon, or to not have it all just fall back on Earth; theorists are still battling with their models to figure it all out, but it seems likely that only a few scenarios with particular impact angles and velocities could result in our Moon. In other words, the formation of our Moon may have been a bit of a fluke, and it's entirely possible that large moons around rocky planets are pretty rare.

'If you were an astronomer living in another planetary system and you were looking at Earth through your telescope, you'd think, "wow, you'd need a protoplanet of exactly this size to smash into this planet at exactly this angle, and then things could happen,"' muses Caltech's Jessie Christiansen.

The Apollo missions themselves helped to cement the impact origin model of the Moon's formation, with the pristine lunar rocks brought back to Earth by the Apollo astronauts proving to bear a remarkable resemblance to the materials that make up Earth's crust and mantle – a fact that makes perfect sense if the Moon is made mostly with debris from Earth's crust and mantle.

Yet the Apollo astronauts were only following in the footsteps of many fictional adventurers who'd visited the Moon before real-life rocketry caught up. In 1865 and 1869, the forefather of science fiction, Jules Verne, wrote his novels *From the Earth to the Moon* and *Around the Moon*, respectively, about a successful mission to Earth's nearest neighbour. (Written, one might add, with remarkable foresight, given that the real Apollo missions were still a century away.)

Verne was not the first to write about visiting the Moon – how could he be, with that silvery disc above us sharing the sky with the stars and tantalizing our imaginations? – and he certainly wasn't the last. Stories by H. G. Wells and Edgar Rice Burroughs followed, and the Moon became a familiar target in the SF pulps of the 1920s, '30s and '40s, before the Space Age ushered in more realistic portrayals. Yet even once we'd been there, the magic never really wore away.

In fact, the magic has spread. Mercury and Venus lack a moon; Mars has two, but they're just tiny lumps of rock. The gas and ice giants Jupiter, Saturn, Uranus and Neptune, however, have 282 moons between them, Saturn leading the way with a remarkable 146 moons, albeit most of these are small moonlets just a few dozen kilometres across. Even tiny Pluto has moons, five of them including Charon, which is half the size of Pluto itself. We also find moons around some asteroids, and even keeping denizens of the Kuiper Belt – a ring of icy bodies far out beyond the orbit of Neptune – company. A moon like Earth's Moon may or may not be rare, but it seems that other types of moon are commonplace.

Now that astronomy has turned its attention to the even more distant worlds of exoplanets, thoughts naturally turn to moons around those exoplanets. These are exomoons, and we've found two decent candidates so far. A warning, though: they are as alien to us and our experience of moons in the solar system as hot Jupiters were to the planets of the solar system when astronomers first began detecting them in the 1990s.

While science struggles to detect exomoons that must surely be there, science fiction has pushed ahead, though perhaps not as eagerly as one might have expected. Perhaps because when people think of moons they often think of our barren Moon, or perhaps SF is waiting for science to fill the breach.

The most famous SF film ever made, *Star Wars* (1977), featured an exomoon. No, not the Death Star – as Ben Kenobi said, 'That's no moon.' The rebel base, however, was on Yavin IV, the fourth moon of a giant planet around which the Death Star had to pass before it had a clear shot with its superlaser. Fortunately, Luke Skywalker blew up the Death Star just in the nick of time. From what we saw of Yavin IV, it was a lush jungle world, very much like Earth (well, Guatemala actually, where those exterior shots were filmed), and a far cry from the cratered, lifeless textures of our Moon.

Two years later, the crew of the commercial space tug *Nostromo* was fatefully awoken by their ship's on-board computer when it detected a distress signal coming from a nearby moon, LV-426. The rest is history, as they inadvertently unleashed the most terrifying alien in the entirety of cinema in Ridley Scott's *Alien* (1979). LV-426 is dark and cold. Interestingly, by the time of James Cameron's sequel, *Aliens* (1986), LV-426 was being 'terraformed' by human colonists, thickening the atmosphere artificially to render it fit for human life. This suggests that LV-426 was at least as large as Mars – large enough to hold on to an atmosphere for a time, at least, though it would have to be constantly replenished as it leaked into space. In the first film we see LV-426 along with two other moons orbiting a Saturn-like giant planet with rings as the *Nostromo* approaches in a wonderful scene that combines physical effects, great direction and an eerie score by Jerry Goldsmith.

The *Star Wars* saga returned to exomoons in *Return of the Jedi* (1983), in which the Empire is building a second Death Star around the forest moon of Endor, home to the cute/annoying Ewoks (delete as applicable; I was young enough at the time to find them endearing). Though we're told it's a moon, we don't see a parent planet at any time, and Endor is quite clearly large enough in its own right to support an Earth-like environment.

The second sequel to Arnold Schwarzenegger's superior SF action film *Predator* (1987), in which his combat team are stalked through a Central American jungle by an alien hunter, appears to be set on an exomoon. Although it is not explicitly stated in *Predators* (2010), we do see what appears to be a Jupiter-like giant planet in the sky and several other moons, including one spewing debris from what looks like a collision. It's kind of an anti-*Earthrise*, a scene of alienness and cosmic violence. And as an aside, the Predator favours warmer climates and views the world in infrared light. One might surmise that the Predator's species (the Yautja) is indigenous to a planet nearer the inner edge of its system's habitable zone, and its sun might perhaps be a red dwarf star, which because of its cooler temperature emits the majority of its light in infrared.

But it was a little film from the year before that really became the poster child for exomoons.

## Pandora

James Cameron's *Avatar* (2009) eased its way into cinematic history with a worldwide box-office take of a mind-blowing $2.9 billion, making it the highest-grossing film of all time. The 2022 sequel (the first of four sequels planned in total), *Avatar: The Way of Water*, wasn't quite as successful, 'only' scoring a worldwide box office take of $2.3 billion and placing it third on the list of films with the highest gross, behind *Avatar* and Marvel's *Avengers: Endgame* (with *Titanic* in fourth place, meaning James Cameron has made three of the top four highest-grossing films of all time, as of 2024).

By putting so many bums on seats, the *Avatar* films have been the most high-profile presentation of exoplanet – or in this case, exomoon – science. As it happens, by inspiring one of the world's leading experts in exomoons, it brought things full circle.

Back in December 2009, when *Avatar* was conquering all at the box office, the astrophysicist René Heller took his wife on a date to see the film. 'She didn't find it all that interesting,' Heller recalls, but for Heller personally the film had a big impact. It wasn't the 3D release or the blue-skinned characters, or the way that the world of Pandora was brought to immersive life. It was the fact that Pandora was an exomoon.

'I hadn't even thought about exomoons until I watched *Avatar*,' he says, 'but the next day I couldn't stop thinking about it.'

Heller's scientifically powered imagination went even deeper than the film did. In the fiction, Pandora is a Mars-sized, Earth-like moon (its gravity is said to be one-third of Earth's, which is why everything, including the native Na'vi, is so tall) orbiting a giant planet called Polyphemus, which in turn orbits alpha Centauri A, a Sun-like star in the alpha Centauri triple system. In reality, alpha Centauri A and B are a wide binary, with Proxima Centauri – discussed in detail in Chapter Six – even further away, and together they are the closest star system to the Sun. Three planets have been found around Proxima, but none have been identified around alpha Centauri A or B – yet.

Anyway, in his musings, Heller realized that there would be eclipses of the sun by the giant Polyphemus that would be routinely visible on Pandora (we see these eclipses in *The Way of Water*). If Pandora is tidally locked to Polyphemus – like Earth's Moon, which always shows to us the same familiar face – then the planet will not appear to move in the sky, but the sun and stars will. If Jake Sully, the main character in the *Avatar* films, played by Sam Worthington, was stood at the sub-planetary point (the location on Pandora where the planet is directly overhead), then at noon, when it is supposed to be the brightest time of day, it would actually be the dimmest because the sun, Alpha Centauri A, would be occulted by Polyphemus.

But Heller also realized that midnight would not be the darkest time of night: 'If Jake Sully were standing at the sub-planetary point at midnight, so the sun was below his feet on the other side of Pandora, Polyphemus would act as a ginormous mirror above, reflecting bright sunlight.' Depending upon Polyphemus's albedo, this could cast Pandora in a bright twilight, despite it being in the middle of the night. Heller started scribbling 'illumination curves', detailing how alpha Centauri A and Polyphemus in tandem would illuminate Pandora, day or night. There would be several peaks in daylight either side of the midday eclipse, and then a smaller peak building up around midnight. In fact, thanks to their parent planet acting as a mirror, exomoons like Pandora would receive more illumination overall than their parent planet does, because they'd be lit up by both the direct light from their star and the reflected light from the planet.

Tidally locked worlds orbiting M-dwarfs, like the planet January in Charlie Jane Anders's *The City in the Middle of the Night*, where the sun never moves, would play havoc with biology's diurnal cycle. In a different way, a moon orbiting a planet like Pandora orbits Polythemus would also experience a disrupted diurnal cycle, with night during day and day during night.

Computer models confirmed that Heller's intuition was spot on, and with Rory Barnes of the University of Washington in Seattle, he went into more detail about how the relationship between a planet and its moons could affect the habitability of those moons. Besides the extra illumination there's also heat radiating from the planet as well as additional tidal heating to consider – that is, the way the gravitational tidal forces emanating from a planet can flex a moon's interior like putty. These tides result from a moon's non-circular orbit around its planet, meaning that a moon feels more gravity from its parent at certain times, leading to friction in its interior, and this friction generates heat. We see tidal heating in action in our own solar system, on the moons of Jupiter. For example, on Jupiter's innermost moon, Io, the gravitational tides churn up the moon's rocky mantle, leading to Io being littered with hundreds of volcanoes that leave the moon covered in lava fields and lakes of molten magma, rendering it completely inhospitable. Meanwhile on Europa, which is the next moon out, Jupiter's gravitational tides have the reverse effect and increase habitability as the heat produced by the tides melts the ice, forming Europa's giant global subterranean ocean while at the same time powering hydrothermal vents on the floor of that ocean. The resulting geochemistry in those vents quite possibly creates the conditions for life.

Neither Io nor Europa are Earth-like. For moons like Pandora it's vital that the tidal heating takes place in a moon's core and not its mantle. This is because tidal heating in the already hot core increases the difference in temperature between the core and the mantle, and this temperature gradient is what gives rise to convection currents that drive both plate tectonics and a magnetic dynamo. In a way, it's analogous to the way hot air rising through convection generates weather systems.

We've covered the importance of plate tectonics and planetary magnetic fields in previous chapters, but it's even more important for a moon to have a strong magnetic field. This is because it doesn't just have to ward off radiation from its sun, but also radiation trapped in its parent planet's radiation belts. Earth has radiation belts, called the Van Allen belts, though our Moon lies far beyond them. Jupiter has them too, but on a much larger scale, befitting Jupiter being a much larger planet. Given that Jupiter's magnetic field is the most powerful of all the planets, it's able to sweep up lots of charged particles from the solar wind (the constant stream of charged particles emanating from the Sun) that flows past the planet. Add to this the Io Plasma Torus – a doughnut-shaped ring of charged particles including ions of sulphur dioxide that originate from the volcanic plumes of Io that spew into space – and all in all, it's a pretty deadly radiation environment. It's a good job that Europa's ocean is covered by dozens of kilometres of solid ice to protect it from this radiation. Notably, magnetic fields on moons are rare in the solar system. Only Jupiter's largest moon, the planet-sized Ganymede, has its own innate magnetic field powered by the dynamo effect.

All of this – the reflected light, the tidal heating, the danger from radiation belts – led Heller and Barnes to conclude that there is a 'habitable edge' around planets.

'Interior to this habitable edge, the sum of all the tidal heating, thermal illumination from the planet and reflected sunlight from the planet is too huge for the moon to be habitable, and if it has water it would experience a runaway greenhouse effect and become uninhabitable,' says Heller.

So, we start to picture what boxes Pandora has to tick in order to be habitable, or 'super-habitable', as Heller describes it, meaning even more habitable than Earth – Pandora fits this description nicely, being lush with life. It was the detailed biology of Pandora's ecosystem that particularly appealed to SF author Stephen Baxter, who wrote *The Science of Avatar* in 2012.

'The makers of Avatar went into vast detail about Pandora and the other bodies in the alpha Centauri system,' says Baxter, who is used to writing fiction based on science, but when writing *The Science of Avatar* had to reverse the process in a way. 'For me it was about looking at what they'd produced and rationalizing it. There were a few things that didn't really bear much scientific scrutiny, such as unobtanium [the fictional mineral that humans were mining on Pandora]. I thought the network of trees and consciousness was far more plausible than this stuff they were digging up to provide energy for the Earth.'

## Detecting exomoons

While there are over 5,700 examples (and counting) of exoplanets on which to base theoretical models of habitability, there's a severe dearth of known exomoons – in fact, none have been confirmed to exist yet. In 2022 Vera Dobos of the Kapteyn Astronomical Institute at the University of Groningen in the Netherlands led a study that modelled 4,140 known exoplanets. Her simulations estimated that of those 4,140 worlds, 234 stood a greater than 1 per cent chance of possessing a habitable moon, and only seventeen had higher than 50 per cent chance. The two best systems in which to look, suggest Dobos and her team, are Kepler-459b (70 per cent) and Kepler-456b (69 per cent), both of which are giant planets on wide orbits in the habitable zones of their star. Crucially, however, no hints of any exomoons have been observed around any of these planets, even if theory predicts that they should be there. In fact, astronomers have only discovered two decent exomoon candidates anywhere, and one of them even Heller disputes. So let's say one-and-a-half good candidates. We'll come on to them and their surprising characteristics later in the chapter. But for now, it says something when the touchstone of our ideas about exomoons is the fictional Pandora, rather than anything factual. As such, it means that we can only take our ideas so far before a lack of data stymies attempts to do anything in greater detail.

'This is why I switched over from the theory of exomoon habitability to actual observations [searching for them], because I had the impression that everything we can say has been said from a theory point of view,' says Heller: 'Now we need those observations to find those moons, and parameterize and characterize them, and feed our equations into them. And then the next step would be to see if they do have water, but that will take decades I think.'

The problem in finding exomoons isn't that we think they will be rare per se, like Earth's Moon, but that finding any exomoon poses a real challenge. You might think that we should be finding more exomoons because astronomers have the ability to detect the transits of planets as small as Mercury, and Ganymede and Saturn's moon Titan are both larger than Mercury. However, it's not that simple. As Jessie Christiansen warns, 'Exomoons are right on the edge of our detectability.'

Despite advances over the past thirty years, our planet-finding methods are still biased towards discovering worlds close to their star. The closer the planet is, the more often it transits, which allows astronomers to build up a good signal that they can be confident in. A planet in the habitable zone of a Sun-like star transits about once per year, so astronomers have to wait years to get enough data.

The planetary systems orbiting red dwarf stars are in scale with their diminutive star, meaning that they orbit close to their star and have modest diameters, being mostly smaller, rocky planets. Giant planets are rare around red dwarfs, with only a handful known. Similarly, the size of an exomoon is expected to scale with the size of its parent planet. So although potentially habitable planets orbiting red dwarf stars make frequent transits because they are so close to their star, their moons are generally going to be too small for our current telescopes to detect. A moon the size of Earth's Moon, which is 3,475 kilometres (2,160 mi.) across, is really at the absolute extreme of what astronomers can currently detect. So it's no surprise we've found no moons in red dwarf systems yet.

What about hot Jupiters? As giant planets they should have larger moons, but research suggests otherwise. If we remember, hot Jupiters form further out from their stars and then migrate inwards. Astronomer Fathi Namouni of Université Côte d'Azur in Nice, France, has shown that as a giant planet migrates inwards over the course of 100,000 years, its moons are pulled away from it by the growing tidal forces from the looming parent star. Namouni's simulations showed that a migrating giant planet with four large moons would see three of them stolen from it and the fourth crash into the planet itself, the ultimate ignominy. These lost moons stripped from hot Jupiters could be ejected by the star onto long, Pluto-like orbits. A team of South American astronomers led by Mario Sucerquia of the Universidad de Antioquia in Colombia suggest that these lost moons might explain some of the giant exocomets and their huge tails that have been detected transiting some stars. Sucerquia's team want to call these objects 'ploonets' – a portmanteau of planet and moon.

One possible scenario that could lead to some moons remaining around hot Jupiters is that of moons colliding rather than being ejected, and new moons arising from the debris once the hot Jupiter settles into its new orbit close to its star. Perhaps this is the scene of destruction presented to the characters when they look skywards in the aforementioned film *Predators*. Overall, though, we shouldn't expect to find moons orbiting hot Jupiters, which is a shame, as they would otherwise make promising targets.

If exomoons are going to be rare around hot Jupiters, and too small to detect around rocky planets, then that leaves giant planets like Jupiter and Saturn that are further from their star. This is where finding them becomes even more difficult.

Jessie Christiansen describes how an exomoon might appear to us in a light curve as a planet transits its star. Recall that a light curve describes how a star's brightness changes with time as a planet moves in front of it, blocking some of its light. Ordinarily, given the round, symmetrical nature of a planet, this light curve is a nice U-shaped dip. The two exomoon candidates discovered thus far were detected because one of the shoulders of the U was lower relative to the other, which is where the moon was peeking out from behind the limb of the planet, creating an additional – minor – dip in starlight. If the moon was behind the planet at the time of transit, we wouldn't see it; if it was too far off to the other side, it would have transited on its own and we might not have recognized it.

'You have to look for a configuration at just the right time where the moon isn't too far away that it transits separately to the planet, and when they're not right on top of each other from our line of sight,' says Christiansen.

The further away a planet is from its star, the more precise the alignment has to be for us to see the planet transiting. And then, when they do transit, we have to wait years to see more transits.

'You need to observe hundreds – perhaps a thousand – stars to see even one planet transiting [at large orbital radii],' says Heller: 'So you need a lot of time and a lot of luck, and you need to observe hundreds of thousands of stars to find even a couple of moons.'

Future missions might redress the balance a bit. The European Space Agency's PLATO (PLAnetary Transits and Oscillations of stars) mission will survey up to a million bright stars over the course of ten years in search of transiting planets and their moons. Our best chance of finding exomoons, however, lies with the Nancy Grace Roman Space Telescope, a wide-field survey instrument that will detect microlensing events – tiny gravitational lenses that temporarily magnify the light of background stars as they move in and then out of alignment. Many of these lenses will be caused by the gravity of unseen foreground planets bending space, but some may also have additional lensing caused by the gravity of exomoons.

The Roman Space Telescope is, at the time of writing, scheduled to launch between October 2026 and May 2027; PLATO should see the stars around the same time. We await the discoveries that they will make with eager fascination.

## Giant exomoons

In principle, a giant planet could have many large moons. With the extra heating that a moon would receive, the giant planet wouldn't necessarily even have to reside with the star's habitable zone for its moons to be habitable. We see that potentially in our own solar system, with Jupiter's moon Europa and Saturn's moons Enceladus and Titan being prime spots to look for life. Though they're not like Earth the way that Pandora is – Polyphemus must be closer in to alpha Centauri A than Jupiter is to the Sun – there's more than one way a world can be habitable.

'*Avatar* reminds us that moons may be habitable things,' says Knicole Colón, deputy project scientist for exoplanet science on the JWST: 'That's always inspiring to me because we find a lot of giant planets, and they could have moons that are habitable, even if the giant planets themselves are not.'

The best candidate moons that we've discovered so far both orbit giant planets larger than Jupiter, and sticking to that scaling law we mentioned earlier, the candidate moons are similarly oversized.

The naming convention for exomoons is a roman numeral – I – after the planet's designation. In *Star Wars*, the rebel base is on the moon Yavin IV, the fourth moon of the giant planet Yavin, so this is basically correct. The two exomoon candidates discovered thus far are catalogued as Kepler 1625b-I and Kepler 1708b-I (of which Jessie Christiansen is a co-discoverer), and they orbit giant planets that are at roughly Earth-like orbital distances from their star. Most bizarre, though, is the size of the moon candidates, with both being of similar mass to Neptune. That's right, they're the size of an ice giant and would dwarf Earth! 'This is another one of those selection biases,' says Christiansen: 'The first moons we should find should be the biggest moons.'

Like hot Jupiters, which were also discovered first, these Neptune-sized exomoons are probably rare in the grand scheme of things. And like hot Jupiters, they challenge a lot of our theories. Though our Moon was formed following an impact, the moons of the giant planets in our solar system formed out of discs of rock, gas and water that accreted around the giant planets when they were young. Astronomers have even imaged such a moon-forming circumplanetary disc around a young exoplanet, PDS-70c, which is a giant planet with twice the mass of Jupiter located 370 light years away from us. However, as large as Ganymede and its compatriots are compared to our Moon or even Mercury, the amount of mass needed to make all of Jupiter's current moons is much less than what models predicted the mass of Jupiter's moon-forming ring to be. So, one suggestion is that Jupiter and Saturn actually used to have more moons, perhaps even larger moons, going through several generations that repeatedly spiralled to their doom into their parent planet where they were swallowed whole, like the whale swallowed Jonah. The moons we see today are the survivors that were last to form. Planetary scientist Robin Canup of the Southwest Research Institute has spent a great deal of time modelling the formation of large moons, both in our solar system and beyond, and her calculations indicate that the total mass of all the moons to form around a planet in this way is 0.0001 times the mass of the parent planet. Therefore, a planet with ten times the mass of Jupiter can form a moon with one-third of the mass of our Earth. The giant moons around Kepler 1625b-1 and Kepler 1708b-1, if they are indeed real, could not have formed in this way because they break that mass ratio between planet and moon. So it's fair to say that moons on this scale were not expected, but that's the beauty of the universe – it likes to surprise us.

Theoretical models can't replicate the formation of these giant moons in orbit around their parent planet. While our Moon doesn't follow Canup's mass ratio either, we know it formed differently, that is, from an impact. The giant exomoon candidates couldn't have formed from an impact because they orbit giant planets with deep, gaseous atmospheres

'Rocky planets can have impacts in that way,' says Christiansen. Gas giants are different. We've seen impacts before on Jupiter, principally the fragments of comet Shoemaker–Levy 9 in 1994

– 'But when Shoemaker–Levy 9 hit Jupiter, the fragments just vanished, poof!'

Anything impacting a gas-rich giant is just going to be swallowed by it to join all those unfortunate early generations of moons in its belly. So we can rule out 1625b-1 or 1708b-1 forming from an impact. All that leaves is the possibility that they were captured by the gravity of their adopted parent planets during the early days of their planetary system, when the planets still hadn't settled down into their orbits. While this also requires a bit of fine-tuning in terms of the two planets' relative velocities and approach angles, each needing to be just right to allow capture and not collision, it's apparently not so strange. We see it in our solar system with Neptune's large moon Triton, thought to be a captured Pluto-like object, and Saturn's outermost moon Phoebe, a captured comet. In fact, Phoebe's nature as a captured body snagged by Saturn's gravity as it hurtled in-system is a key plot point in James S. A. Corey's series *The Expanse*, starting with the first book *Leviathan Wakes* (which forms season one and the first half of season two of the TV adaptation). In the story, Phoebe is sent towards our solar system by mysterious ancient aliens who intend to use the Moon to deliver a biological payload onto Earth. This biological 'bomb' involves what the characters refer to as the 'protomolecule', which is capable of taking over all life that it encounters and turning it into something more like the aliens. Only in this case, their plans are thwarted when Saturn accidentally gets in the way and catches Phoebe. Of course, meddling humans discover the protomolecule on Phoebe and set about a chain of events that reverberates throughout the nine-book series. Suffice to say, while we have visited Phoebe with a spacecraft – NASA's Cassini mission flew past it in 2004 and took pictures before entering into orbit around Saturn – we've not discovered any alien biology on it.

But back to the giant exomoons. In fact, it might be more accurate to think of them as double planets, since the common centre of mass around which they both orbit is likely to be somewhere in the space between moon and planet, rather than inside the parent planet. So the 'moon' cannot be said to be orbiting the planet. Once more, we do have examples of this in our solar system, such as the dwarf planet Pluto and its largest companion, Charon.

'Kepler 1625b-I and 1708b-I are probably more like binary planets than they are a planet and a moon,' says Christiansen, who despite participating in the discovery of Kepler 1708b-I is keen to emphasize that these are currently only candidate exomoons. They require verification, which would come from a confirmation of their mass to show that they are a real object and not some spurious background noise. This mass measurement would involve a brilliant technique that takes advantage of what astronomers call 'transit timing variations', or TTVs for short. They've been observed for exoplanets that orbit close to one another, such as the seven worlds of TRAPPIST-1, and could in principle be used to confirm exomoons. The idea is thus: when two planets are close enough their gravity acts on each other, pulling them backwards or forwards depending upon where they are in their orbits relative to each other. In pulling each other around, they change the time that they transit from that predicted for them if the only gravitational force they felt was from their parent star. The amount by which the timing of their transit varies from prediction tells astronomers the mass of the other planet that's pulling on them. In this way astronomers can discover planets even if they aren't seen to transit, simply through the way they influence those worlds that are seen to transit. While astronomers discover exoplanets through TTVs, there's no reason it can't be employed to measure the mass of moons tugging on planets too. With luck a mission such as PLATO will stand a good chance of making advances in this area.

Until then there's always the next three planned *Avatar* films to keep exomoons in our thoughts. Or maybe Disney will make a sequel to the little-known Ewoks TV movies *Caravan of Courage* (1984) and *Ewoks: Battle for Endor* (1985), which were both straight-to-video fare with nary a stormtrooper to be seen. Of course, Endor also featured in a segment of 2019's *Rise of Skywalker*, where we saw a new part of the moon, a stormy sea into which part of the destroyed Death Star II had crashed.

Or maybe we don't have to wait for fiction to experience alien moons. As Colón reminds us, we've got some right here in our own solar system, and they are amazing and ripe for exploration in both fiction and fact. Europa, the quintessential ocean world, is probably our best chance for finding life elsewhere in the solar system. It's unclear how thick its crust of surface ice is, but estimates predict several dozen kilometres. Astrobiologists test technologies for drilling, or melting, through the ice to the ocean underneath by exploring some of the approximately four hundred subglacial lakes beneath ancient ice in Antarctica. Some of these lakes have been sealed off, in darkness, for millions of years: we've already encountered the largest, the 250-kilometre-long (155 mi.) Lake Vostok, in Chapter Five; it hasn't seen daylight for at least 15 million years, possibly longer. Any life in the ocean, cut off from the outside world, would have evolved along its own trajectory, but the trick to exploring Lake Vostok and any life it holds is learning how to do so without contaminating its pristine environment. A recent project to drill into another of the subglacial lakes, namely Lake Whillans, which is located 800 metres (2,625 ft) beneath the West Antarctic Ice Sheet, found the lake to be teeming with life, the project identifying about

4,000 different microbial species through DNA sequencing. The lifeforms are methanotrophs, which employ enzymes that allow the microbes to use oxygen to oxidize methane molecules, stripping electrons from the molecules that the microbes can use as a biochemical source of energy. The amount of available energy and organic carbon in the lake suggests that it could host larger life-forms than microbes.

A similar process could possibly support life in Europa's ocean. So far, no space mission has landed on Europa, though several have flown past and studied the icy moon from Jovian orbit. Currently, two new missions are headed out there: the European Space Agency's JUICE (Jupiter ICy moons Explorer) blasted off in April 2023 to arrive at Jupiter in 2031; NASA's Europa Clipper set sail in late 2024 on a speedier trajectory to reach Jupiter by 2030. Again, neither will land on Europa – despite the experiments in Antarctica, the technology required to explore the moon's ocean remains some way off.

We have landed on Saturn's giant moon Titan, the only moon in the solar system to have an atmosphere. It's not breathable, though, instead filled with hydrocarbons (molecules formed of hydrogen and carbon atoms). When the European Space Agency's Huygens probe parachuted down to Titan's surface in 2005, it imaged terrain covered in rivers of liquid hydrocarbons such as methane and ethane, rivers that feed into large lakes of the same stuff. Though too cold for life as we know it, with surface temperatures as chilly as −174 degrees Celsius, Titan also plays host to a sub-surface ocean like Europa's, which could potentially be habitable. So too could the ocean inside Titan's fellow moon Enceladus, which squirts geysers of water vapour from that ocean into space through deep cracks in the surface opened by tidal forces emanating from its parent planet of Saturn.

Intriguingly, science fiction hasn't exploited these fascinating moons on our own doorstep as often as perhaps it could. Paul McAuley's 2008 novel *The Quiet War* is set on the moons of Jupiter and Saturn. The 1981 film *Outland*, starring Sean Connery, is set on Io, where Connery plays a cop at a mining outpost on the moon – but considering that it was released only two years after Voyager 1 and 2 discovered the volcanoes on Io, this new information doesn't seem to have been incorporated much into the film. Ben Bova featured the exploration of Titan in his 2006 novel of the same name, which formed part of his long-running *Grand Tour* series that featured stories set on all the worlds of the solar system as humans explored, and sometimes fought over, them. The 'found footage'-style film *Europa Report* (2013) documents a doomed crewed mission to Europa that discovers life in the moon's ocean. Stephen Baxter also wrote a novel called *Titan* (1997) about a similarly doomed crewed expedition, but to Saturn's largest moon instead. The story ends billions of years into the future, where an ageing Sun is beginning to expand into a red giant, warming the outer solar system and rendering Titan hospitable to life. This is the ultimate fate of the Sun in about 5 billion years' time, but in Arthur C. Clarke's *2010: Odyssey Two* (1982), which is a sequel to *2001: A Space Odyssey* (1968), the monolith-building aliens act to increase the mass of Jupiter with myriad self-replicating machines in the form of monoliths, to the point where Jupiter's mass grows so great that nuclear fusion ignites in Jupiter's core and it becomes a modest M-dwarf star designed to breathe life on Europa, with one famous, final message for humanity: 'All these worlds are yours – except Europa. Attempt no landing there.' With the help of a monolith, Europan life evolves there tens of thousands of years into the future.

Colón thinks that the interest in exomoons from the *Avatar* series could inspire greater interest in our solar system's own wonderful moons. 'I hope we're going to be more inspired because NASA's funding a mission to Titan that's going to involve a robotic helicopter called *Dragonfly* flying around, which is amazing,' she says. 'I think it might reinvigorate interest in the possibility of moons having oceans. Imagine living on a moon like that and having a gigantic Jupiter-sized planet next to you.'

We thought that *Earthrise* was beautiful, but imagine the scenario depicted by René Heller earlier in this chapter. Imagine being a member of a civilization on the far side of a tidally locked moon. Imagine being the first to go exploring around the other hemisphere, perhaps taking to the sea to cross a vast ocean, on a voyage evoking Ferdinand Magellan's on Earth. And imagine, as you near the other hemisphere, seeing a giant planet rise above the horizon in front of you. Nothing could beat the grandeur of planet-rise, as a huge world adorned in writhing cloud belts and swirling storms, the size of small planets themselves, breaks the horizon, all so close that you feel you could almost touch them. It's science fiction for now, but perhaps somewhere else in the galaxy, far, far away, it's a beautiful fact of nature.

# 9

# WEIRD WORLDS

Sometimes truth is stranger than fiction. The very first exoplanets to be discovered were bizarre beyond belief. Nobody expected there to be planets orbiting pulsars, born out of the wreckage of exploded stars, and yet there they were. Hot Jupiters, too, completely confounded expectation: astronomers were convinced that exoplanetary systems would look like ours, with smaller rocky planets closer to their sun, and giant planets further out. When the first exoplanet to be found around a Sun-like star, 51 Pegasi b, got the ball rolling, astronomers were astounded. What was a giant planet doing there, just 7.78 million kilometres from its star? Conventional wisdom stated (and still does) that gas-rich giants must form at a greater distance from their star. Of course, we now know that hot Jupiters did form further out and migrated in-system, but in 1995 their existence was a shock to us and a reminder that nature can sometimes be stranger than we can possibly imagine.

Or perhaps not. Were there any precedents to pulsar planets and hot Jupiters in the pages of science fiction? Possibly. While there are no direct one-to-one analogues with hot Jupiters, there are fictional exoplanets that start to venture down that road. Goldblatt's World, in Larry Niven's 1984 novel *The Integral Trees* and its 1987 sequel, *The Smoke Ring*, combines both the concept of hot Jupiters and planets orbiting neutron stars (as we shall see, pulsars are fast-rotating neutron stars).

Goldblatt's World orbits just beyond a neutron star's Roche limit. The Roche limit is the point of no return. As long as an object orbits outside the Roche limit, it is safe, but the moment it steps over that invisible line gravitational tidal forces will act to tear and twist the object apart, be it a comet, a moon or a planet. Astronomers have seen this in action before, right here in our own solar system. In 1992 comet Shoemaker–Levy 9 passed inside the Roche limit of Jupiter, coming within 40,000 kilometres of Jupiter's tumultuous cloud tops. That's a remarkably close fly-by. Jupiter's gravity tore the 1.8-kilometre (1 mi.) comet into 21 individual fragments, all following each other on the same trajectory, looking in images like pearls on a necklace. On their next passage, in July 1994, they got even closer to Jupiter – too close, in fact: all 21 fragments slammed into the giant planet. The resulting fireball plumes from the impacts in Jupiter's atmosphere were visible to the Hubble Space Telescope. It was incredible to watch the cometary fragments smash into the giant planet one after the other – a reminder of the ferocious power of nature.

Anyway, the Roche limit can vary from object to object depending upon the density of both the primary body and of the orbiting body. Jupiter's Roche limit is at a distance of about 70,000 kilometres, depending on the other object. Jupiter's innermost moon, Io, orbits at a distance of 422,000 kilometres, well outside the Roche limit. However, one compelling theory is that Saturn's rings were formed when a large moon or comet entered Saturn's Roche limit and was ripped apart some 100 million years ago.

As mentioned, a neutron star is born from a supernova, which is the explosion of a star more than eight times as massive as our Sun. Stars consume hydrogen in nuclear reactions to produce the energy that lights them up. When they run out of hydrogen in their core, they start fusing helium instead. In stars like our

Sun, this is where things stop. More massive stars, however, have high enough temperature and pressure in their core to continue fusing heavier and heavier elements, each element the by-product of the nuclear reactions that preceded it. This process is called stellar nucleosynthesis, and its realization came from the West Yorkshire-born astrophysicist and SF author Fred Hoyle, along with astronomers Margaret and Geoffrey Burbidge and William Fowler, in 1954. Hoyle's most famous SF novels are probably *The Black Cloud* (1957) and *A For Andromeda* (1962, co-authored with John Elliot).

Stellar nucleosynthesis continues creating new elements until iron is produced in a massive star's core. Since stars cannot fuse iron, because the reaction requires more energy to be put in than comes out, fusion in a massive star stops at this point. Without a source of energy to hold the star up, gravity takes the advantage and causes the inner regions of the star to collapse, forming a neutron star, while the infalling outer layers bounce off it and explode outwards, producing the supernova. It's a frightening amount of energy that gets released – a supernova can shine brighter than its host galaxy – but the neutron star survives the conflagration.

Neutron stars are highly magnetic, and early in life they can shoot magnetically collimated beams of charged particles from their poles. Because they are born spinning, we see these beams flashing our way like a cosmic lighthouse, and hence this is what we call a pulsar.

So, returning to the topic of Goldblatt's World, Niven describes it as orbiting an old neutron star that's spun itself down and is no longer a pulsar. As a neutron star it is incredibly dense, as great as $8 \times 10^{17}$ kilograms ($17.6 \times 10^{\wedge 17}$ lb) per cubic metre – a spoonful of neutron-star material would weigh as much as Mount Everest. In comparison, Goldblatt's World is – or rather was – a low-density gas giant with a rocky core. The disparity of densities means that the Roche limit of a neutron star, where a gas giant like Goldblatt's World is concerned, is about 2 million kilometres, assuming that the planet has a density similar to Jupiter, which is 1.33 grams (1⁄20 oz) per cubic centimetre. By any definition, a gas giant planet just 2 million kilometres from its star is a hot Jupiter. That said, Goldblatt's World could not have been that close to its parent star when it formed, otherwise it would have been utterly destroyed by the explosive blast as its star went supernova. However, the destruction of the star and the loss of all its mass will have sent gravitational fields into flux and may have caused Goldblatt's World to migrate inwards.

Being located just outside the Roche limit has some dramatic effects on Goldblatt's World. The 2.5-Earth-mass rocky core of the planet is able to hold together, albeit as an inhospitable and tidally locked world, but its light and fluffy atmosphere has no chance. The neutron star's gravitational tides grab at the planet's gaseous envelope, tearing it away. The gas forms a ring – the so-called Smoke Ring – a million kilometres wide around the neutron star. In Niven's books, life is able to inhabit this Smoke Ring, including the eponymous Integral Trees, which are a kind of plant that grows up to 100 kilometres (62 mi.) in length, their tip pointed towards the neutron star in a kind of tidal-locking scenario.

'This ring of gas fills the whole orbit around the neutron star, so you've got this big ball of rock and endless sky and plants and people who live in this endless sky,' notes St Andrews's Emma Puranen, who if we remember from earlier chapters is studying trends between fictional exoplanets and how they relate to real exoplanet discoveries. It's safe to say that nobody has yet discovered anything like Goldblatt's World with its Smoke Ring, but '[Niven] worked out the maths to get the gravity right at different points in the sky. It's fascinating,' says Puranen: 'I think it's a really creative world.'

We know that planets can be whittled down to their core in real life, as with mini-Neptunes that are turned into water worlds when they migrate too close to their star. Astronomers have actually seen the stripping of planetary atmospheres in action many times, but the mechanism is different to the tidal forces that Goldblatt's World is subjected to. Rather, astronomers have been amazed to discover evaporating planets too close to their stars, the planets' atmosphere being blown away by a fierce stellar wind. The atmospheric material then streams away in a long tail, as if following a planet-sized comet.

Niven's tales of a world around a neutron star beat the discovery of pulsar planets by almost a decade. As surprising as it was for astronomers to discover planets around dead stars, it wasn't completely unexpected. There had been several false alarms before bona fide pulsar planets were officially discovered in 1992 by radio astronomers Dale Frail and Aleksander Wolszczan, who found two planets around the pulsar PSR B1257+12, which is located 2,300 light years from Earth.

The pulsar planets were not discovered by any of the regular methods, such as transits or radial velocity measurements. Pulsars are quite unique objects. Highly magnetized, they are born spinning fast. Their magnetically funnelled beams emit radio waves that our radio telescopes can detect, so the pulsing that we observe as beams flashing in our direction, much like the light emitted from a lighthouse, is actually at radio wavelengths (the orbiting planets might also see pulses in visible light). These pulses are incredibly regular, so regular in fact that pulsars are among the best timekeepers in the universe. And as previously mentioned, they spin incredibly fast – PSR B1257+12 rotates a dizzying 161 times per second, or once every 6.22 milliseconds. Though pulsars are small, maybe 20 kilometres (12½ mi.) across, they are quite literally a blur, and any discrepancies in the timing of their pulses very quickly becomes apparent. This was the case for PSR B1257+12, which had timing anomalies caused by the gravity of two orbiting exoplanets (later found to be three exoplanets) pulling on the pulsar and incurring a Doppler shift in the pulsar's radio beams.

Alastair Reynolds featured a system containing a neutron star as well as a regular star and several planets in his first novel, *Revelation Space* (2000), and its sequels. Its inclusion wasn't actually inspired by the discovery of the pulsar planets; while studying for his degree in the 1980s at the University of Newcastle, he became fascinated by neutron stars and their incredible density.

'I'd been thinking about what you could do with really dense matter and whether you could use it to make a computational matrix out of a neutron star,' he recalls. 'That was the big idea I had and I was trying to build a short story around that.'

That idea ultimately grew into *Revelation Space*, launching Reynolds's writing career. The story revolves around the mysterious extinction of an alien species whose relics are found on the planet Resurgam, and an artificial object orbiting Hades, which is a neutron star in the same system. The artificial object, called Cerberus, was built long ago by aliens to, as Reynolds puts it, 'mess with us'.

In the book, Hades and Resurgam orbit the star delta Pavonis. Delta Pavonis is a real star, a smidgen shy of 20 light years from Earth and found in the Southern Hemisphere constellation of Pavo, the Peacock. As far as we know, there is not really a neutron star in the delta Pavonis system, and no planets have yet been discovered.

'I didn't know any actual neutron stars that I could use that are close enough for the characters to get to with relativistic space travel,' says Reynolds. The closest known neutron star is actually 400 light years away, but admittedly, once they have spun themselves down and stopped pulsing, they can be very hard to detect. 'So I just took a known star – delta Pavonis, though it was Arcturus in the first draft – and put a neutron star in orbit around it. I just thought that if the separation between Hades and delta Pavonis was wide enough, then it couldn't be excluded by our actual observations of delta Pavonis. But I made a bit of a booboo: in the original draft their separation was something like 10,000 AU [one astronomical unit, AU, is the average distance between Earth and the Sun, which is 149.6 million kilometres], but in a later draft I shortened it because I needed the characters to get from the planet [Resurgam] to the neutron star in a shorter timeframe. Now, with the knowledge that we have of stellar kinematics, we can rule out any possibility that delta Pavonis has a neutron star that close.'

Only half-a-dozen pulsar planets have been discovered so far. To be honest, they are not considered an important target for most exoplanet hunters, which might explain why so few have been found. No one is entirely clear how they form, but it seems probable that they are created after a supernova explosion, from either debris ejected by the supernova itself or material made available when a companion star to the pulsar crosses its Roche limit and is torn apart. Some of the pulsar planets themselves are weird and wonderful beyond the simple improbability of their existence. One planet, orbiting the pulsar PSR B1620-26, resides 6,000 light years away within the globular star cluster Messier 4 and is the oldest known exoplanet, with an age of 12.6 billion years – just 1.2 billion years younger than the universe itself. Another, orbiting the pulsar PSR J1719-1438, located 4,000 light years away, is a hot super-Earth that has probably had its thick atmosphere stripped away, perhaps not too dissimilarly to Goldblatt's World, and which contains more carbon than oxygen, leading to its composition being rich in carbon-dominated materials such as graphite and diamond.

## Bejewelled worlds

Carbon-rich worlds are potentially a commonplace phenomenon. Of course, there is carbon on Earth – we are even sometimes described in science fiction as carbon-based lifeforms – but on our planet, there is more oxygen than carbon. We're not necessarily talking about the oxygen in the atmosphere, but the oxygen that's bonded with other elements, such as silicon, to form silicates, which are rock-forming minerals that constitute the bulk of our planet's composition.

On a planet with a higher ratio of carbon to oxygen (C/O), however, the composition would be different. Rather than silicates, the bulk of a carbon-rich planet would be made of minerals such as silicon carbide and titanium carbide, and even a deep layer of pure diamond many kilometres thick. On the surface would be mountains of graphite, and volcanoes would spew chunks of diamond into the air.

In the brilliant TV series *Babylon 5*, the planet Minbar, which is the homeworld of one of the show's principal alien species, the Minbari, is described as being rich in deposits of crystal and diamond. From this material the homes and temples of their beautiful cities are directly cut, carved and etched onto. In one second-season episode, the Minbari ambassador, Delenn, describes how during the spring, the patterns of colour caused by sunlight refracting through the crystal are breath-taking. We don't see much of Minbar during *Babylon 5*'s five-season-long story, but one of the establishing shots of Minbar that we do get shows a towering waterfall cascading between two buildings carved into the diamond, and the planet itself is described as being the seventh planet from its sun – probably nearer the outer edge of its habitable zone, since a quarter of Minbar is covered by a north polar ice cap.

However, in reality, waterfalls and ice caps are not something that we can expect to find on carbon-rich diamond planets. Water is, of course, partly made from oxygen atoms, but the problem is that on carbon planets any oxygen will react with the carbon, becoming locked up in molecules such as carbon dioxide and carbon monoxide, while hydrogen, which makes up the other part of water molecules, will wind up trapped in hydrocarbons such as methane and even more complex molecules. There would be no water, and a carbon-rich planet would be completely inhospitable to life as we know it. Certainly the Minbari, who appear rather human-like – aside from a bone-like crest around their heads – and who breathe air and drink water, wouldn't be able to survive on Minbar, in any case.

Another example of a carbon-rich planet can be found in Jessie Christiansen's favourite SF series, *Doctor Who*: in this case there are 'waterfalls', but of sapphires. 'The episode "Midnight" [2008] sees David Tennant's Doctor and a few different people trapped in a shuttle on this planet called Midnight, and the reason it's cool is that it's a crystal planet,' says the Caltech astronomer.

Then, in 2012, astronomers announced that they had identified an exoplanet, 55 Cancri e, with a high C/O ratio. The planet is in a wide binary system that features a primary star, a little smaller and cooler than our Sun, and a red dwarf. Five planets are known to orbit the primary, including planet 'e', which is a tidally locked super-Earth with a mass at least eight times that of our Earth. In fact, it was the first super-Earth ever found around a Sun-like star, in 2004. Modelling of the interior of the planet by planetary scientists led by Nikku Madhusudhan, then at Yale University but now at the University of Cambridge, determined that 55 Cancri e must have a carbon-rich interior, with up to one-third of the planet's mass being composed of carbon. Deep in 55 Cancri e's interior, the pressure would compress that carbon into diamond. NASA's Spitzer Space Telescope, which observed the universe in infrared light, detected variations in surface temperature on 55 Cancri e that have been attributed to violent volcanism and expansive lava plains partially hidden by clouds of volcanic dust, while the Hubble Space Telescope found no evidence of water vapour in 55 Cancri e's atmosphere, indirectly supporting the carbon-rich prediction.

'Scientists realized that there is probably crystalline carbon, basically diamonds, in the crust of 55 Cancri e,' says Christiansen. 'That was really cool because, for a hot second, I thought, they've found Midnight! Eventually other people measured the C/O ratio and decided that it wasn't that extreme, though the people who originally claimed it still stand behind their analysis.'

So 55 Cancri e may or may not be like Minbar and Midnight. However, planets with a high C/O ratio undoubtedly exist out there, somewhere, given that we find stars with similar high ratios. Our best chance of confirming one will be the JWST, which can spectrally analyse planetary atmospheres and measure the C/O ratio directly and with high precision.

## Heavy metal planets

Planets aren't just made up from silicate rock or carbon; there's plenty of metal in there too. We've seen how iron forms in the core of a massive star that then goes boom; more metals, such as aluminium, are formed in the ferocious temperatures and pressures of the supernova explosion itself, or in the merger of two neutron stars, an event called a kilonova that produces large quantities of some of the most precious metals, such as gold and platinum. These heavy elements then wind up being present in the protoplanetary clouds that give rise to planetary systems, and thus in planets like Earth, sinking through the molten interior to congregate in a planet's core. What metals we mine from the surface of our planet were brought here by subsequent asteroid impacts.

We can't access Earth's metallic core, of course, except for in Jules Verne's imagination (specifically his 1864 novel *Journey to the Centre of the Earth*) or in bad Hollywood science (the 2003 film *The Core*, directed by Jon Amiel), but early in the solar system's life the planetary environment was a violent place, with protoplanets smashing into each other left, right and centre. Some of the remains of this planetary demolition derby survive to this day, found among the denizens of the Asteroid Belt between Mars and Jupiter. In particular, there are chunks of planetary core in the form of the metallic M-type asteroids, the largest of which is called Psyche. In October 2023 NASA launched a mission, simply called Psyche, to this ancient relic to learn more about the cores of planets.

Presumably exoplanets form in the same way as the planets of our solar system, with metallic cores. But how much of a planet can be made from metal? Mercury, the closest planet to the Sun in our solar system, is 4,880 kilometres (3,032 mi.) across, but

4,040 kilometres (2,510 mi.) of that is made up of its metallic core. It's not clear why its metallic core is so large comparatively, but the leading explanation is that early in Mercury's life, the innermost planet was whacked by a giant impact that stripped away much of Mercury's crust and mantle.

Could entirely metal planets exist? The Transformers, a favourite of mine ever since childhood, come from the metallic planet of Cybertron, which seems logical given they are robotic sentient beings. The first issue of Marvel Comics' *The Transformers* (1984) describes Cybertron as being as large as Saturn (the second-largest planet in our solar system) and that robotic life arose there thanks to 'naturally occurring gears, levers and pulleys' – an ingenious if not exactly realistic idea. Later in the series, writer Simon Furman retconned the origin of Cybertron and the Transformers into a somewhat more mystical story, imagining a war of the gods between the good Primus, and his army of Light Gods, and the evil Unicron. Primus sought to end the conflict by trapping his and Unicron's essences inside large asteroids, and this is where it gets interesting because Primus then reshaped his asteroid into the metal planet Cybertron. Though now smaller than the claim of Saturn-sized from the first issue of the comic, a large asteroid or protoplanet does seem more in keeping with Cybertron's appearance, where cities, buildings and bridges are clearly visible in great detail from orbit. Relevant to this discussion is the fact that some asteroids are metal-rich pieces of planetary core. I'd be willing to bet that Primus sculpted Cybertron out of an M-type asteroid. And Unicron? He turned his asteroid into a gigantic robotic body that could transform into a planet with a gaping maw that ate other planets. Because why not?

If Cybertron had truly been a sphere of metal the size of Saturn, which is 116,460 kilometres (72,365 mi.) across, then its surface gravity would have been truly great. It's a mistake to think that a planet's gravity scales equally with mass. For example, you might think that a rocky super-Earth with five times the mass of our Earth would have five times the surface gravity, but it does not. This is because we have to factor in the distance of the surface from the centre of the planet, that is, the radius. A five-Earth-mass rocky planet has a radius about 1.5 times as large as Earth, and by dividing the mass by the square of the radius, we can calculate the surface gravity to be 2.2 times as strong.

So let's calculate the surface gravity of a Saturn-sized Cybertron. For simplicity, let's assume that it's made of pure aluminium – a sphere of aluminium with a radius of 58,230 kilometres (36,182 mi., nine times larger than Earth's radius) and a mass of $2.21 \times 10^{28}$ kilograms, which is 3,701 times the mass of Earth. To calculate surface gravity relative to Earth, we can simply divide Saturn-sized Cybertron's mass in Earth masses (3,701) by the square of its radius in Earth radii ($9.1^2$), arriving at a surface gravity 44.7 times greater than on Earth. It doesn't matter how strong Cybertron's warring inhabitants, Optimus Prime's Autobots and Megatron's Decepticons, are – nobody is moving around under that kind of gravity. An asteroid- or protoplanetary-sized Cybertron makes much more sense.

Hal Clement's novel *Mission of Gravity* (1954) takes planetary surface gravity even more seriously. Famed for the rigorous depictions of science in his stories, Clement had a degree in astronomy from Harvard University. In *Mission of Gravity*, a human-built robotic probe becomes stranded at the north pole of the planet Mesklin, but Mesklin is no ordinary planet. Clement referred to it as a 'whirligig' planet because it spins incredibly fast, once every eighteen minutes – so fast, in fact, that the planet is flattened into a highly oblate spheroid with a centrifugal bulge.

Centrifugal force is a form of inertia that pushes outwards from a rotating body. To simulate gravity on space stations, it has been suggested by scientists and engineers such as the late Gerard K. O'Neill of Princeton University that cylindrical, hollow space stations could be spun up, with the 'ground' on the interior wall of the cylinder and the centrifugal force pushing people and things towards or against that 'ground'. We can see this in the rotating space-station wheel in *2001: A Space Odyssey*, or on board the 5-mile-long (8 km) cylindrical space station in *Babylon 5*, where centrifugal force provides not just gravity but the outward inertia to launch the station's Starfury fighters.

Back to Mesklin: as an oblate spheroid, it's far wider at the equator (Clement calculated 77,250 kilometres (48,000 mi.)) than at the poles (31,770 kilometres (19,740 mi.)). As such, the gravity of this sixteen-Jupiter-mass planet is far stronger at the poles – which are closer to Mesklin's centre of mass – than it is at the equator. In *Mission of Gravity*, Clement calculated that at the equator (where there would be a sharpish edge around the planet as a result of the hectic spin), the surface gravity would be three times Earth's gravity, but at the poles, it would be a whopping 665 times Earth's gravity. No wonder the probe gets stuck there. After the fact, Clement recalculated the gravity at the poles to be 'only' 225 times Earth's surface gravity. Regardless, the point remains that the gravity is supremely strong at Mesklin's poles. Under such conditions the terrain would be mostly flat, mountains or even hills unable to lift themselves up in the intense gravity. In the end, the probe is retrieved by one of Mesklin's intelligent centipede-like inhabitants following a dangerous journey, in exchange for the humans' advanced scientific knowledge.

'Mesklin is an extremely exotic planet that was questionably physically possible even by the standards of the time and completely ludicrous now,' comments the exoplanet scientist and SF podcaster Alex Howe of NASA's Goddard Space Flight Center. 'But the way that Clement goes into analysing it and how it would work with that extreme gravity field was a really nice piece of writing.'

## Worlds reshaped by gravity

In fiction, Superman's homeworld of Krypton is somewhat similar to Mesklin in the sense that Krypton has a higher gravity than Earth, which is the science used to explain some of Clark Kent's superpowers, which manifest in Earth's lower gravitational field. The real planet that comes closest to Mesklin in shape is WASP-12b (if anybody is wondering, its name comes from the Wide-Angle Search for Planets). It's a tidally locked hot Jupiter and is described as being 'egg-shaped', not because of rotation, like Mesklin, but because of the gravitational tidal forces from its parent star that are stretching the planet at the equator. WASP-12b is also dark – well, pitch black, really – since it reflects a paltry 6 per cent of the light incident upon it; the rest of the light is absorbed by the planet's atmosphere, which contains oxides of titanium and potassium. An even blacker planet is TrES-2b (also known as Kepler-1b), which reflects less than 1 per cent of light that shines on it. In fiction, the darkest planet can be found in another Hal Clement novel, *Close to Critical* (1964), which features human scientists conducting remote exploration of the planet Tenebra. Again, rotation plays a significant role, but this time it's the star rather than the planet. Tenebra orbits Altair, which is a real and bright star that is obvious in the summer night sky as the most southerly of the stellar trio that form the Summer Triangle. Altair, which is 16.7 light years away from Earth, is known to rotate rapidly in real life, spinning once in just under eight hours, leading to it becoming flattened at the poles. In *Close to Critical*, Clement describes how more light is radiated from Altair's poles than the equator because the poles are closer to the star's core nuclear engine. Since in the fiction Tenebra's orbit is tilted with respect to Altair's equator, at some points in the planet's orbit it is over the star's poles and experiences summer as it receives more light and heat, and plunges into winter while level with the star's equator. But that's not the most interesting thing. In the novel Tenebra has no clouds, just a thick, diffuse smog that absorbs all of Altair's light incident upon it. Clement came up with this bizarre environment where the atmosphere is so thick at the surface that it becomes buoyant, with raindrops up to 15 metres (50 ft) wide gently floating down through an atmosphere with a pressure eight hundred times the surface pressure on Earth. Because it is so dark, plant life uses chemical energy to exist, like the black smokers in vents deep in Earth's ocean, rather than photosynthesis.

Before we move on from the stories of Hal Clement, there are two more things worth noting about Mesklin. The first is that at sixteen times the mass of Jupiter, Mesklin would be what astronomers call a brown dwarf, a kind of failed star: they're not quite large enough to ignite nuclear reactions of hydrogen in their core (although the larger brown dwarfs can burn deuterium for a short while), and they are thought to form in similar fashion to a star. Brown dwarfs are also gaseous; a solid planet of such a mass certainly wouldn't form naturally.

The second thing to note is that Clement placed Mesklin in the 61 Cygni star system because at the time astronomers had thought they'd found evidence for an exoplanet candidate around 61 Cygni. That turned out to be false, and despite intensive searches, no planets have yet been identified there. However, 61 Cygni is also famous for being the first star to have its distance revealed, by Friedrich Wilhelm Bessel in 1838; it is 11.4 light years away.

Gravity has also reshaped the worlds of science fiction in other ways. In Robert Forward's 1984 novel *Flight of the Dragonfly*, also known as *Rocheworld*, the crew of a lightsail-driven spacecraft called *Dragonfly* travel to the red dwarf Barnard's Star, one of the closest stars to the Sun, just 5.96 light years away. The existence of a real planet orbiting Barnard's Star was announced in October 2024, with hints of three other planets also present in the radial velocity data. The confirmed planet, Barnard b, is one of the most diminutive ever found with about one-third the mass of Earth, and orbits too close to Barnard's Star to be habitable. The other possible worlds, if they exist, are similarly small. There's a long litany of claimed discoveries of planets orbiting Barnard's Star that were eventually refuted, but finally astronomers think they've found the real deal.

If you can get past the clunky dialogue, dated characterization and plodding plot, there's a lot of smart thinking in Forward's book, notably regarding his imagined planet around Barnard's Star – or more accurately, planets. For this world is a contact binary, a double planet whose atmospheres are actually touching, forming a bridge between them. One of the planets is rocky but dry, the other covered in ocean laced with ammonia, and the gravity of each planet acts upon the other, deforming them into egg shapes.

Astronomers know of stars that are contact binaries, but they are objects a million kilometres across or more, with large, distended, overlapping atmospheres and bridges of matter being shared from one to the other. The only plausible double world in our solar system is Pluto and its largest moon, Charon, which is half the size and one-tenth of the mass of Pluto.

## Exotrojans

No contact exoplanet binaries are known, at least not yet, but astronomers have found evidence for planets that share the same orbit. Again, this phenomenon has previously been seen in our solar system.

Lagrangian points are gravitationally neutral areas where the gravitational pull of a planet and the Sun balance, allowing a smaller object such as an asteroid or a spacecraft to remain in these areas relatively unperturbed. Indeed, space agencies send spacecraft such as the JWST to Earth's second Lagrange point, L2, which is 1.6 million kilometres away in the opposite direction to the Sun.

Two more Lagrange points are found 60 degrees in front of and behind a planet in its orbit. Jupiter, which has the biggest gravitational field of all the planets in the solar system, has been able to corral at least 9,800 asteroids into these two Lagrange points. Astronomers call them trojan asteroids, and they've also been found in smaller numbers in the Lagrangian points along Mars's, Uranus's and Neptune's orbits. Two have even been found in Earth's L4 Lagrange point, which precedes Earth in its orbit around the Sun by 60 degrees.

In 2023, astronomers spotted what could be the first exotrojan, in the very young planetary system known as PDS 70, which is 37 light years away. It is an incredible system; at just 5.4 million years old, it is our best look at a system in which planets are still forming around their star. From our vantage point we see PDS 70 almost face-on. With the help of two telescopes in Chile, namely the European Southern Observatory's Very Large

Telescope and its SPHERE (Spectro-Polarimetric High-contrast Exoplanet Research) instrument, as well as ALMA (the Atacama Large Millimeter/submillimeter Array), astronomers have actually imaged the rings of dust in the planet-forming disc around the star, and two planets carving gaps in that disc to form those rings. The planets are giant planets, though they appear as little more than fuzzy blobs in the images. Nevertheless, at least one of these proto-Jupiters, PDS 70c, appears to be forming an exomoon from a circumplanetary disc that has coalesced around the planet, while PDS 70b shows more tentative evidence for such a disc. But PDS 70b also has something more: a faint object shadowing 60 degrees behind it at the L5 point in its orbit. Whatever the object is, it is faint, possibly surrounded by a dust cloud, and might currently have the mass of two of Earth's moons. So if it *is* real, it's tiny compared to its sibling planets – perhaps PDS 70b swept up the vast majority of gas and dust in its orbit, leaving this exotrojan to make do with whatever was left.

That's one exotrojan; how many planets could, in theory, share the same orbit? 'There's an episode of *Doctor Who* ['The Stolen Earth' (2008)] where they show up in the TARDIS and there's 27 planets all in front of them and I'm like, "that's not dynamically stable!"' laughs Jessie Christiansen.

Perhaps not 27 planets, but 24 planets all jostling for position might be a different kettle of fish. It sounds like science fiction, but a team of astronomers led by Sean Raymond of the University of Bordeaux in France have shown that such a system could be stable for billions of years. Employing computer simulations, they modelled different numbers of planets in the same orbit, all the way up to 24. In many of the configurations the planets remain equally spaced, but sometimes they are able to swap places via what we call horseshoe orbits.

Don't think of these 24 planets as pearls on a string, following perfectly one after another. The cosmos is never this organized! Rather, they'll have slightly different orbital radii, and slightly different eccentricities, meaning that some planets will follow a slightly more elliptical path than others. This results in them being able to dip in and out of each other's orbits. While in many configurations they remain equally spaced, in others this dipping in and out enables them to swap places with each other. We see this in our own solar system, not with planets but with two of Saturn's small moons, Janus and Epimetheus. Astronomers call them shepherd moons because they shepherd one of Saturn's rings, their gravity keeping the edge of the ring finely defined, and as they do so, they swap positions in this co-orbital fashion.

How 24 planets on the same orbit could form naturally is unclear. Raymond's team suggest that were such a system ever to be discovered, it could be evidence for technological extraterrestrial life because it's hard to imagine how such a system could be anything but artificially assembled (though how one might move planets is another matter entirely). However, it's feasible that smaller numbers of planets might end up co-orbiting naturally, as evidenced by the possible sign of exotrojans.

Not all planets have to necessarily orbit the same star all of the time. The University of Birmingham's Amaury Triaud, who if we remember specializes in tidally locked worlds and circumbinary planets, has developed the concept of the 'bouncing planet'.

In short, his model describes how an exoplanet can bounce between two stars in a binary system. One of the stars has two planets orbiting it, which begin exchanging orbital energy in the form of angular momentum. This causes the innermost planet to shorten its orbit, while the outermost planet increases its orbit.

'Eventually, that might bring the outer planet to the other star, and it ends up bouncing between the two stars,' says Triaud: 'You could imagine a civilization on such a world, with their planet able to travel to the nearest star.'

Although no such bouncing planets have ever been discovered, Triaud's model shows that in principle, it could work. Alas, such a system would not be stable forever. Over long timescales – hundreds of millions, or billions, of years – a bouncing planet will either be ejected or crash into one of the stars, posing an existential risk to the hypothetical civilization living on it. 'I think it would be a pretty good setting for a SF book!' enthuses Triaud.

## Rogue planets

That's all the weird orbits, but what about exoplanets that make no orbits at all? Indeed, such worlds do exist. Astronomers have counted several hundred free-floating nomad planets with no star to call home – but of course, SF got there first.

The classic 1951 SF film *When Worlds Collide* features a rogue star colliding with Earth, but the film is actually based on Edwin Balmer and Philip Wylie's novel of the same name from 1933. In the novel, two rogue planets enter our solar system, with one of them destroying Earth in a collision and the other replacing Earth around the Sun, providing a new home for the survivors of humanity.

The novel *When Worlds Collide* partly inspired Alex Raymond's *Flash Gordon* comic strip a year later, with the rogue planet Mongo, ruled by the villainous Ming the Merciless, on a collision course with Earth (the 1980 film starring Sam Jones and a very entertaining Brian Blessed changed this backstory and had Max von Sydow's Emperor Ming want to destroy Earth just because he felt like it). In the first issue of Marvel Comics' *The Transformers*, in 1984, Cybertron is shaken free from its orbit by the sheer violence of the war between the Autobots and the Decepticons and is sent careening across interstellar space as a rogue planet, only to be endangered by our Asteroid Belt as it entered our solar system, 4 million years ago. (As an aside, fiction usually portrays asteroid belts as being densely packed with chunks of rock – the TIE fighters chasing the *Millennium Falcon* in *The Empire Strikes Back* is one example, as is the Obi-Wan Kenobi/Jango Fett spaceship chase from the *Star Wars* prequel *Attack of the Clones* (2002). In reality, there are huge spaces between the asteroids in the asteroid belt, and if you were standing on one asteroid, you'd be unlikely even to see another asteroid for most of the time.) Gerry Anderson's *Space: 1999* TV series, which ran from 1975 to 1977, kind of featured a rogue planet in the sense that a nuclear accident miraculously blasts Earth's Moon into interstellar space. And the homeworld of the Founders in *Star Trek: Deep Space Nine* is a rogue world hidden inside the fictional Omarion Nebula.

Star-forming nebulae and young star clusters are actually great locations in which to find rogue, or free-floating, exoplanets. For example, in 2021 astronomers announced the discovery of a large collection of rogue planets, finding a bounty of at least seventy, and possibly as many as 170, free-floating Jupiter-mass planets in the Upper Scorpius OB Association. An OB association is a collection of many young O- and B-type stars (if we remember Annie Jump Cannon's OBAFGKM stellar classification system that we discussed in Chapter Five, O- and B-type stars are the hottest and brightest) that have only recently formed, and they are often found still within or near their birthing nebula. The rogue planets, which are thousands of times fainter than the neighbouring stars, either formed like stars themselves (blurring the lines between stars and planets) by condensing directly out of the nebula, or were stripped from their parent stars by gravitational tidal forces wielded by the other stars in the tightly packed association. Assuming that the number of gas-giant free-floating planets found in the Upper Scorpius OB Association is typical, then astronomers estimate that there could be billions of Jupiter-sized planets wandering lost amid the space lanes of our Milky Way galaxy, and even more Earth-mass rocky rogue planets, the small size of which makes them much more difficult to detect. Indeed, estimates of the total amount of material typically lost from a young planetary system simply as a result of migrating giant planets bowling everything out of their way like skittles are as great as twenty Earth masses. That's a lot of planetary material chucked into open space.

More recently, in 2023, new research led by Sam Pearson of the European Space Agency found another treasure trove of potentially 540 rogue exoplanets in the Trapezium star cluster in the heart of the famous Orion Nebula. The discovery was made with the JWST, and most are huge gas giants; the smallest seems to be about two-thirds the mass of Jupiter. But most bizarrely, forty of these objects were found as wide binaries. Pearson calls them JuMBOs: Jupiter Mass Binary Objects. Did they form directly from the nebula like binary stars do, or were they somehow ejected in tandem from their planetary system? The jury is still out.

At least one rocky rogue planet has been detected: a world designated OGLE-2016-BLG-1928, with a mass about the same as Mars, spotted by the Optical Gravitational Lensing Experiment (OGLE) that looks for random gravitational microlensing events. The search for rogue planets will be one of the science drivers for NASA's forthcoming Nancy Grace Roman Telescope, which, as discussed in the previous chapter, will be able to detect planetary-scale microlensing events en masse.

When it comes to the universe, weirdness seemingly is part of the territory. Who knows what strange new world we will find next, or which unconventional but scientifically sound idea will debut in SF to blow our minds anew. Bring it on!

# 10

# ECUMENOPOLISES

Space opera has its particular traditions, furnishings that mark the story out as rip-roaring high adventure across a universe filled with imagination. Ray guns? Check. Spaceships? Check. Conquering bug-eyed aliens? Double check. And for SF author Arkady Martine, one other quintessential facet of space opera is city planets, one of the most well-known being the world city of Coruscant, which is the primary seat of power in the *Star Wars* universe and is completely covered by one incredibly large conurbation. 'City-planets are one of those things that's just in the water,' she says.

Martine should know – besides being trained in city planning and working in infrastructure and energy policy in real life, her debut novel *A Memory Called Empire* stormed the Hugo Awards (the literary world's leading SF awards) in 2020. Martine's story features a world-city – an 'ecumenopolis' – at the heart of her Teixcalaan interstellar empire where a young ambassador is caught between scheming politicians secretly vying to usurp a wilting emperor.

As you read this, there's a good chance that you, dear reader, are living in a town or a city. After all, in England in the year 2021, some 82.9 per cent (56.3 million) of you did, an increase of 6.2 per cent since the previous census in 2011. In Scotland, 4.97 million people live in an urban settlement, with only 491,330 living in the countryside. It's a similar story for the rest of the UK. In

Europe, 74.8 per cent of the population live in towns and cities. In the United States, 80 per cent of the population are city dwellers. Across the world as a whole, more than half of the planet's population is urbanized, and that number is rapidly increasing. The UN's Population Division estimate that by the year 2050, this figure will have risen to 68 per cent, and to 85 per cent by the dawn of the twenty-second century.

We're increasingly becoming a species of city dwellers as we carelessly eradicate green belt, hack away at rainforests and pave over ecological reserves. More than that, we're not just a planet of big cities; we're heading towards becoming a planet of megacities, as urban areas merge into large metropolises. An ecumenopolis just takes that idea several steps further.

The concept of an ecumenopolis is fictional and theoretical, for now, but why shouldn't we consider it? In lieu of any exoplanets with actual known ecumenopolises, we can compare instead to the urbanization of our Earth. Furthermore, it's not beyond reason that if there is life out there on another planet that has developed society and technology, then they could have constructed huge urban areas or even artificial cities or planets. Could we detect an ecumenopolis from Earth with our telescopes? Ecumenopolises are worth studying, even if for no other reason than to appraise us of the possibilities and warn us of the challenges and risks that a world-city poses.

Ecumenopolises may be a major trapping of space opera, but the wider genre of science fiction has always had a close relationship with the ideas and concepts built into our cities. Think of Los Angeles in the year 2019 in *Blade Runner* (1982), neon lit and under oppressive and seemingly constant night; the huge skyscrapers of the wealthy in Fritz Lang's *Metropolis* (1927); the crime-ridden depths of Neo-Tokyo in the famed Japanese anime *Akira* (1988);

or Mega-City One, where Judge Dredd maintains law and order in his inimitable draconian style.

'My references also include William Gibson's "Sprawl" as well, which I like to think of as a city-planet in progress,' says Martine. The Sprawl refers to a fictional metropolis on the northeastern seaboard of the United States, encompassing the Boston and Atlanta areas, in cyberpunk author Gibson's *Sprawl* trilogy, featuring his famed 1984 novel *Neuromancer*, plus *Count Zero* (1986) and *Mona Lisa Overdrive* (1988).

One cannot wander around a large city such as London, Paris or New York and fail to recognize aspects of those fictional cities in the real ones. In London, try walking down Charing Cross Road on a dark, busy evening, the bright neon lights of shops, the mix of ethnicities, the aroma rising from street-food vendors: you feel palpably like Harrison Ford as he moved through night-time futuristic LA looking for clues to the whereabouts of the escaped replicants in *Blade Runner*. All that's missing is the flying spinners.

*Blade Runner*, *Neuromancer*, *Judge Dredd*; the cities in these stories are just examples of single mega-cities – enormous, unwieldy, vulnerable to crime and inequality, but single cities nonetheless, on a planet of many separate cities. If you carry on walking southeast out of London, for example, you'll eventually find yourself in the Kent countryside. There is an end to the city.

Imagine, though, that you didn't arrive in the countryside. Imagine that you just kept walking, and you kept finding more and more of the city, a city that never ended, a city that was the entire world. This, like Coruscant, or the Jewel of the World that is the city-planet at the heart of Martine's Teixcalaan empire, is the essence of an ecumenopolis: a single urban area covering the entire surface of a world.

'Ecumenopolis' is a word that was coined in 1967 by the influential urban planner Constantinos Doxiadis, who held the doctrine that the science of human settlements favoured larger and larger urban areas, culminating in a city-planet.

'Doxiadis was very optimistic about growth,' says Juan Miguel Kanai, a human geographer at the University of Sheffield who has written about the theory of ecumenopolises and their portrayal in science fiction. 'Back in the sixties, society was optimistic about growth. Today we don't think about growth in the same way.'

Arkady Martine concurs and expands on the different mindset back then: 'I think it's interesting to contextualize ecumenopolises in the 1960s, because the sixties were really the height of ideas that had been presented by people such as [the pioneer of modern architecture] Le Corbusier, who were attentive to a very administrative approach as to what patterns people should live in – patterns of a city, of buildings, of architecture, of transit, of open space.'

However, Martine feels that SF has become a little too locked-in on depicting just one kind of metropolis. 'As science-fiction writers we're not very good at using SF to think about living in extremely densely populated communities without immediately going to *Blade Runner* as a reference,' she says. 'The vision of ecumenopolises in science fiction is a little bit stuck because it imagines a kind of rigidity, such as dividing areas by social class or function, the idea of people living in very narrow buildings and never leaving them – that's super Le Corbusier, he would have loved that – but that is not reflected in actual cities. The way people use space is a lot less regimented than what people thought they did with it.'

We don't have to look far in science fiction to find what Martine is talking about. The Mega-Blocks in *Judge Dredd*, for example, housing hundreds of thousands of people in vast buildings hundreds of storeys tall, are practically cities in their own right. On Coruscant, which is a world of skyscrapers, the elite live at the tops of buildings, while down below, underneath all the lines of hover-car traffic, is the grime and crime – the real colour of civilization. In Alastair Reynolds's *Chasm City* (2001), the eponymous city, built into a crater spewing volcanic gases, has a similar scenario, with the city divided into two distinct areas: the aristocrats at the top of the skyscrapers, in the part of the city referred to as 'the Canopy', and the shanty town of 'the Mulch' far below.

Such scenarios are ripe for fantastical storytelling that pits the haves against the have-nots. And, of course, there's a hint of real-life flavour to it, with our own cities so divided by inequalities. But the point that Martine is making is that real cities are far more fluid that the rigid sociological strata implemented by authority in what are essentially dystopias. In a real city, a poor street can be found next to a rich street. Luxurious shops mix with the low-rent flats above them. The transit routes are filled with people from all backgrounds. And people make use of the space provided to them in ways that often surprise the planners who want everything in its place.

## The first city-planet

Doxiadis invented the term 'ecumenopolis', but he didn't invent the idea; it was present in SF long before 1967. Trantor, the central planet of a great human interstellar empire in Isaac Asimov's *Foundation* (made up of five connected short stories and novellas, the earliest written in 1942, and published in combined form in 1951) and its sequel novels, is the archetypal example.

'*Foundation* is an allegory for the Roman Empire,' says exoplanet scientist Alex Howe: 'Trantor is Rome, and ancient Rome was this massively oversized city for its time, just as Trantor is at the heart of a massively oversized civilization.'

As we may remember, Howe is an astronomer at NASA's Goddard Space Flight Center, where he studies exoplanets, but he is also a budding science-fiction author and a podcaster, producing several series of his *A Readers' History of Science Fiction* podcast. It was while recording the podcast, and talking about Asimov, among other writers, that Howe became interested in city-planets.

In Asimov's *Foundation* novels (and the modern TV adaptation starring the always excellent British actor Jared Harris), the fall of this huge empire is foreseen by a genius mathematician named Hari Seldon, who is one of science fiction's most celebrated characters, up there with Flash Gordon, Ellen Ripley, Captain Kirk and Luke Skywalker. Seldon uses a fictional concept that Asimov named 'psychohistory', a science that allows the user to probabilistically predict the future. The eponymous Foundation is Seldon's failsafe to the traumatic events that he sees lying in wait in the future of the Galactic Empire, a means of preserving knowledge and expertise on a distant, secret world while the rest of the galaxy burns and falls into disarray. Trantor, being the centre of the Galactic Empire, becomes symptomatic of the malaise that is spreading through the empire and is eventually sacked by rebel forces during a civil war – just like Rome was sacked by the Visigoths in AD 410, the moment that many hail as the beginning of the end of the Roman Empire, at least as a force to be reckoned with.

As an aside of astronomical interest, Asimov originally positioned Trantor both figuratively and literally at the centre of our galaxy. However, this was before astronomers knew about Sagittarius A*, the supermassive black hole 4.1 million times more massive than our Sun that lurks there, surrounded by dense clusters of stars that are prone to going supernova, all resulting in intense radiation fields that would render any planets close by uninhabitable. So in later stories, Asimov described Trantor as being as near to the galactic centre as was safe – in practice, probably a few thousand light years from Sagittarius A*.

Trantor became the inspiration for a host of other ecumenopolises in SF. In Frank Herbert's *Dune*, the homeworld of the villainous Harkonnens is Giedi Prime, a city-planet that is dirty, dark and polluted. We've already mentioned the most famous ecumenopolis in SF besides Trantor, which is, of course, Coruscant. This city-planet debuted in *Heir to the Empire* (1991), the first of Timothy Zahn's superb Lucasfilm-sanctioned sequel trilogy of novels to the original *Star Wars* films (which began a long series of novels that formed an 'expanded universe'; alas, these have now been relegated to an alternative universe by the actual filmed sequel series). Coruscant then cameoed at the end of the 1997 special edition of *Return of the Jedi* before playing more of a starring role in the prequel trilogy films *The Phantom Menace* (1999), *Attack of the Clones* (2002) and *Revenge of the Sith* (2005). From orbit, we see Coruscant criss-crossed with structures, lines and circles of light on the nightside, every square metre of the surface covered in city. In an episode of Disney's *Star Wars* series *The Mandalorian* (2019–), a visit to Coruscant finds a small fenced-off boulder. It's a tourist attraction, a landmark that is actually the peak of Coruscant's tallest natural mountain, named Umate. The mountain is part of the Manarai mountain range, which has been completely covered in city structures (this substantially retcons the landscape in Zahn's *Heir to the Empire*, in which the Manarai mountains are uncovered, deliberately left untouched to provide a picturesque venue for the rich and the elite to visit).

'In one of the chapters of *The Mandalorian*, they referred to Coruscant as an ecumenopolis,' points out Miguel Kanai. 'They actually used the term.'

## The heat is on

So, in theory, is an ecumenopolis a realistic prospect, or would its ecological footprint be far too great to support itself?

I asked the question, and the responses were mixed. Arkady Martine was the most positive.

'I find an ecumenopolis theoretically plausible,' she says. 'Unlikely, but theoretically plausible. It would be hard.'

Miguel Kanai is more sceptical, arguing that we'd really need *Star Wars* technology to build a *Star Wars* planet: 'For a world like Coruscant to work on Earth, we'd have to make an enormous leap of imagination, extrapolating existing technologies to a point that just doesn't seem feasible . . . Without vegetation, oxygen and food have to be manufactured somehow, and it's just impossible with current technology. I mean, we cannot even produce enough fresh water on Earth, let alone figure out how we'd produce air and all of that stuff on an ecumenopolis.'

Alex Howe sees one major stumbling block that could prevent the concept of ecumenopolises from ever getting off the ground, and that's the problem of heat. As anyone who has lived in a city can attest to, cities in the height of summer are able to retain uncomfortably high levels of heat. When I lived in Kent, the thunderstorms from all that heat slowly coming off London and the surrounding boroughs were enormous. According to Howe, in a fictional ecumenopolis it's not climate change per se that is the source of the problem (although it certainly doesn't help), but rather the unavoidability of the Second Law of Thermodynamics, which states that in any system that is doing work, some of the energy will always be converted to waste heat that is radiated into the environment around that system, since heat flows from hotter to cooler regions until everything is the same temperature. This waste heat will build up cumulatively until the city-planet grows so hot that the inhabitants of the ecumenopolis will be cooked.

It was Larry Niven – he of so many of the smart, imaginative, fictional planets that we discuss in this book – whom Howe credits with clueing him in to the issue of heat. One of the alien species in Niven's long-running *Known Space* series is the Pierson's Puppeteers, so named after the human who first encountered them. In Niven's Hugo-, Locus- and Nebula-award-winning novel *Ringworld* (1970), the Puppeteers are escaping the galaxy. One of the Puppeteers, Nessus, describes to the story's main antagonist, Louis Wu, how long ago the Puppeteer population grew, first to half a trillion, then a trillion. All those Puppeteers packed onto one planet, an ecumenopolis, produced lots of waste heat and their planet, Hearth, started to overheat. So they purchased technology from another alien species that allowed them to actually move their planet further from their star to help cool things down. As an ecumenopolis, Hearth lacked farmland, so the Puppeteers have four farming worlds, and they ship their food in from those planets.

When the time came to escape the Milky Way altogether following a chain reaction of supernovae in the core of the galaxy, the Puppeteers manoeuvred their homeworld and their four farming worlds into a Klemperer Rosette – five planets, each at the point of an imaginary pentagon and all orbiting a common point with the same angular momentum. This gravitational system was devised by the German aerospace engineer Wolfgang Klemperer in 1962 (he had six bodies in his original concept, not five), just eight years before the publication of Niven's novel, and it is entirely stable, in theory. In Niven's novel, the Puppeteer's rosette of planets has set sail for the Large Magellanic Cloud, which is a nearby dwarf galaxy about 169,000 light years away. So not only do the Puppeteers have an example of a world city, but they have formed an artificial system of trojan planets, just as we discussed in Chapter Nine. The Puppeteer's rosette is also featured in Niven and Edward Lerner's *Fleet of Worlds* series (2007–12): five books that act as four prequels and one sequel to *Ringworld*.

'The Puppeteer's Hearth is one of the things that keyed me in to the waste heat problem,' says Howe, who is a big fan of Niven: 'I describe Larry Niven as re-inventing hard science fiction [hard, in this context, refers to the rigorous use of real science in a story, as opposed to the "soft" space fantasy of *Star Wars*, for example] not as SF that conforms strictly to known physics, but as SF that invents new physics, or perhaps extrapolates from what we currently know, but applies it rigorously.'

Heat, of course, is a problem on modern-day Earth too. Rising temperatures are being driven by anthropogenic climate change, but we don't help ourselves by the way we construct our cities. Concrete, and especially asphalt, are super absorbers of heat. They hold on to that heat and radiate it back very slowly. So cities assembled from roads and pavements and concrete buildings are naturally going to grow hotter, particularly at the height of summer. We can see this in heat maps, where cities appear to glow several degrees hotter than the surroundings. This is called the heat-island effect.

'The heat-island effect is something that is talked about a lot in cities, but it's more a function of the land-use distribution than just the size of the city,' says Miguel Kanai: 'It's worse in cities that grow larger without enough vegetation and enough green and blue spaces that could have a moderating effect on temperatures.'

The summer of 2022 saw record-breaking temperatures hit the UK, as they did in many parts of the world, at their worst causing a risk to life. In July 2022 temperatures peaked in several locations in the UK, including in London, at over 40 degrees Celsius. A thermal map of the capital showed that the coolest areas, by over 5 degrees Celsius, were the big London parks and commons, such as Hyde Park, Regent's Park, Greenwich Park and Primrose Hill. As Kanai says, smart land use is crucial to keep temperatures down in any city, be it London or Coruscant and its trillion or so inhabitants. Some cities such as Tokyo and São Paulo have severe problems with heat simply because they have so little green space.

There are measures that can be taken to cool our cities – for example, planting trees on the pavement that provide shade, and don't absorb heat as efficiently as tarmac.

'You can also paint roofs with reflective paint, usually very light in colour, and this can reduce the internal temperature of the building without air conditioning by something like 3 to 5 degrees, which is super-impressive because it seems so simple,' says Martine. 'So there are definitely ways in which city planners, construction engineers and architects understand how to make this better, which makes me think that the very real problem that *Ringworld* posits is also one that a civilization or a smart city could work through.'

The effort to make cities greener is why Alex Howe thinks that an ecumenopolis would have to be the greenest city ever seen, with hanging gardens, light-coloured buildings and solar panels everywhere. That latter part is important; nuclear fusion would produce too much heat, he reckons, and so he favours an ecumenopolis run on solar power. If Earth were an ecumenopolis, how many people could live in it? Howe has done the calculations.

He assumes solar panels with 50 per cent efficiency, which is a little higher than today's best lab-tested panels. He also proposes that each person would require 36,000 watts at minimum, which is the sum of each person's energy consumption plus the energy required to grow grain and livestock for food – the inhabitants of Howe's imagined ecumenopolis do not appear to be vegan. Howe suggests doubling this to 72,000 watts to permit any over-indulgences in lifestyle. So, based on this, Howe calculates that by turning Earth into an ecumenopolis it could support 1.7 trillion people. It sounds like a lot, but the population density would be 11,500 people per square kilometre on dry land, which is more than is currently the case for London and New York, but less than the population densities of cities such as Paris and Mumbai.

For a science-fictional ecumenopolis, there may be science-fictional solutions to the heat problem too. Howe imagines enormous radiator fins thousands of kilometres long protruding from city-planets and using space as a heat sink. This isn't too different in principle to the heat radiators on the International Space Station in orbit above Earth, or the six giant, heat-radiating fins on the fictional *Babylon 5* space station. The huge radiator fins climbing up from a city-planet's surface would need to be designed so that they somehow do not absorb heat on the dayside, but radiate the waste heat from the planet into space on the nightside. This design might therefore potentially be better suited to a city on a tidally locked planet – one could envisage an enormous system ferrying excess heat onto the nightside where it's radiated into the coldness of space.

If building giant radiators is difficult, then perhaps there's a more natural solution. 'There's been a paper recently talking about mountaintops as natural radiators,' says Howe. This would be problematic for Coruscant, which as we've seen in *The Mandalorian* has built over all its mountains.

## The city limits

With an eye towards this, Martine's own ecumenopolis doesn't feature buildings on every scrap of land. The Jewel of the World has managed open spaces, agriculture and oceans – it's not a sphere covered in skyscrapers. 'I really wanted to think about what a successful city-planet might look like, and by successful I do not mean utopian, just a city that is not hard to live in.'

Indeed, in most cases, the ecumenopolises of SF tend to not have to deal with the problems of heat, water, food, air, a lack of a carbon–silicate cycle, energy generation and so on because these are problems that they must have solved technologically, otherwise the ecumenopolises wouldn't exist. Even Asimov later introduced farming worlds to supply Trantor, just as Niven did for the Puppeteer's Hearth.

However, from our vantage point on twenty-first-century Earth, it is difficult to comprehend how an ecumenopolis could succeed. Urbanization is certainly a contributor to the increasing environmental damage we are inflicting upon the planet. Take global warming, for example: according to the intergovernmental Organization for Economic Cooperation and Development (OECD), 70 per cent of global energy-related (that is, from power stations and so on) carbon-dioxide emissions come as a result of cities. Pollution, too, is a factor that is enhanced by cities – just think of all that traffic contributing to smog. Greener technologies could improve things, with electric cars, renewable power and the probability that widespread nuclear fusion is now surely only a matter of time, but it's still hard to look past the ecological footprint of cities. And in a despairing example of negative feedback, climate change and the ecological disasters that it brings simply fuel more urbanization.

'In a time of increasing instability in weather- and therefore climate-related disasters, people tend to congregate around infrastructure because most people come to live near other people so that they can help each other,' says Martine: 'Cities are going to become larger and denser, more people are going to live within them within the next fifty years, and cities will have to grow and change to support that.'

There may be limits. In 1972 the so-called Club of Rome (a non-profit organization of academics, business leaders, politicians and civil servants) published their somewhat controversial Limits to Growth report. This posited that continued economic and population growth would begin to butt up against the finite resources of planet Earth, at best stymying growth, at worst leading to the collapse of civilization. It was in stark contrast to the ethos of previous decades, where growth was aspired to. Technofuturists begged to differ, pointing out that technology could help solve many of the problems highlighted by the Limits to Growth report. Genetic modifications helping food production, the shift from coal and oil to renewable energy to remove the reliance on dirty finite resources, the increased market for electric vehicles, a move to put industry in space (the first experiments to beam solar power down to Earth via microwaves took place in early 2023), the continued development of nuclear fusion that does not produce radioactive waste and so on could all help alleviate these claimed limits.

However, simply looking around the world today shows us that it's not that easy. We haven't moved away from fossil fuels yet, an act of stupidity fuelled by the greed of a few to the detriment of everyone. Despite repeated and consistent pleas to reduce emissions and hit a target of global warming of just 1.5 degrees Celsius, we're on course to shoot past that and bring about severe consequences. In 2009 a multinational team of scientists led by Johan Rockström, a professor in earth system science at the University of Potsdam in Germany and a professor of global sustainability at Stockholm University in Sweden, defined nine planetary boundaries for which going beyond the limits would create tipping points heralding ecological and environmental disaster. These nine boundaries are worsening climate change, the depletion of the ozone layer, the acidification of the oceans, global freshwater use, changes in how land is used, the increasing dissolution of the biosphere, chemical pollution, aerosols pumped into the atmosphere and biogeochemical interruptions in the biological nitrogen and phosphorus cycles.

In 2009 Rockström's team of 28 scientists concluded that humankind had gone beyond three of the nine boundaries; by 2015 that had increased to four boundaries being crossed, and as of 2024 six of the boundaries (biosphere integrity, climate change, biogeochemical flows in the nitrogen and phosphorus cycles, the lack of fresh water, the detrimental change in land-use and chemical pollution) have been breached. Cities are not the cause of this per se, but the way we live in them is a huge contributor. The only thing that will save us is changing how we live and, in the process, changing our cities, making them greener and cooler. It's up to us.

## Alien lights

Earth's problems, however, are not necessarily universal. Perhaps if there are technological, urbanized aliens out there, they might not be quite so quick to hit the self-destruct button. Perhaps they have cities that are more in tune with nature, or at least have developed technology that alleviates the problems we now face. If there are large alien cities on exoplanets, or even a fully fledged ecumenopolis, could we detect them?

Let's start with city lights. In 2011 the astrophysicists Avi Loeb of Harvard University and Edwin Turner of Princeton University did a thought experiment. If there was a Tokyo-sized city on Pluto, which orbits the Sun between 39.3 and 39.6 astronomical units (5.88 billion kilometres and 5.92 billion kilometres), could we detect the lights of that city with our current technology? Loeb and Turner think that we could. They don't really think there's a city on Pluto – to silence all doubt, we know there isn't, since NASA's New Horizons spacecraft flew past Pluto and imaged it in great detail in 2015 – but the method with which we could detect such a city could be used on more distant exoplanets. Loeb and Turner point out that as an exoplanet orbits its star, we see the planet at different phases, like the Moon. Depending on the phase, we might see more artificial light from the nightside. This would change the usual light curve we'd expect to see from an exoplanet, and beg the question, why is it appearing brighter than it should at night? Detecting such an anomaly would be a way of revealing significant levels of urbanization. Loeb has more recently re-run the calculations for the JWST, showing that it could detect LED city lights on nearby exoplanets such as Proxima b if the artificial lights on the nightside amount to 5 per cent of the brightness of the dayside reflecting the Sun's light (in which case the aliens would have a serious light pollution problem, as Earth's lights at night are only 0.01 per cent as bright as the sunlit dayside).

Similar research by Thomas Beatty of Steward Observatory at the University of Arizona has shown that a next-generation space telescope, planned by NASA for launch sometime in the 2040s, with an 8.4-metre (27½ ft) mirror – which is 2 metres (6½ ft) larger than JWST's primary mirror, meaning this next-gen telescope can gather much more light – would have an even easier time detecting city lights on exoplanets. It would ably detect artificial lighting twelve times that of Earth on Proxima b, and spot potential ecumenopolises on worlds found in up to fifty nearby star systems.

City-planets have their limits, though. Technological species with particularly advanced engineering skills might decide to forgo their old ball of rock entirely and build a brand new world – a megastructure. In science fiction, Olaf Stapledon first discussed such things in passing in his influential 1937 novel *Star Maker*, which inspired the great twentieth-century physicist Freeman Dyson in 1960 to come up with the concept of what quickly became known as 'Dyson spheres'. He pictured a great shell around the Sun, assembled from raw materials obtained by dismantling the planets in the solar system, most notably the most massive world, Jupiter. This shell would absorb all light emitted by the Sun, and the interior surface could even be lived on. A Dyson sphere that extends as far from the Sun as Earth, which is a distance of approximately 150 million kilometres, would have an interior surface area of 283,000 trillion square kilometres.

It's no surprise that Dyson spheres became popular objects in SF. Perhaps the most famous depiction is in *Star Trek: The Next Generation*'s sixth-season episode 'Relics' from 1992, which is an episode also notable for featuring Scotty from the original *Star Trek*. The engineer is found, preserved and suspended in the transporter buffer of USS *Jenolan*, which had crashed 75 years earlier on the surface of an enormous Dyson sphere described by Data as being 200 million kilometres (124 million mi.) across and having a surface area equivalent to 250 million 'class M' planets (*Star Trek*'s way of saying rocky planets in the habitable zone). When the *Enterprise* becomes trapped inside the Dyson sphere, it's up to Scotty and his counterpart on the *Enterprise-D*, Geordi La Forge, to get the *Enterprise* out.

However, there is a problem with such depictions of Dyson spheres. A giant shell would be inherently dynamically unstable; all it would take is a little gravitational push from a passing star to twist itself apart or even cause it to collide with the hidden star at its heart. This is also a problem for Larry Niven's *Ringworld*, in which the eponymous world is an artificial construct, a giant ring around a star that stretches to a diameter about equal to Earth's orbit around the Sun, on which the interior star-facing surface is filled with oceans and mountains and continents, while giant 'shadow squares' overhead, interior to the ring, regularly block the star's light to bring about night in the area cast in shade. Remember, the Klemperer rosette is in the same novel – *Ringworld* is a book of big ideas! But as a solid structure the eponymous Ringworld is prone to the same catastrophic gravitational forces and tides that a complete Dyson sphere would be.

Instead, Freeman Dyson realized that an enormous and dense spherical swarm of solar collectors, each with their own thrusters for station-keeping, would be preferable to a solid structure, and so perhaps a better name that has been used for Dyson's concept is a Dyson swarm. Each solar collector might still be the size of a planet, with substantial living space.

In 'Relics', the *Enterprise* is unable to detect the Dyson sphere until it is practically right on top of it. In reality, any megastructures – solid structures or swarms – absorbing all that solar energy are going to heat up pretty quickly, and just like an ecumenopolis, they would need to re-radiate away that excess heat in the form of thermal infrared radiation. While we wouldn't be able to see the star inside the swarm, this waste heat should make a Dyson swarm glow in infrared light.

There have been several astronomical surveys searching for the infrared emission from Dyson swarms. Earth's atmosphere – or more specifically, the water vapour within the atmosphere – likes to absorb many infrared wavelengths, so infrared studies of the universe are best done from orbit, as is the case with the JWST. One of the first infrared space telescopes was the not-very-imaginatively named InfraRed Astronomical Satellite (IRAS), launched in January 1983. It actually discovered some of the strongest evidence for the existence of exoplanets prior to their verified discovery in the 1990s: IRAS found dusty planet-forming discs ringing several stars, including Vega, which is just 25 light years away and is one of the three bright stars of the Summer Triangle, along with Altair, mentioned in the previous chapter, which dominates the night sky on warm evenings in the Northern Hemisphere. Vega was just one of 350,000 infrared sources catalogued by IRAS, and astronomer Richard Carrigan of FermiLab near Chicago perused this catalogue in search of anything that revealed emission akin to a Dyson swarm. The problem is that how we think the thermal infrared signature of a Dyson swarm will appear is very similar to how a menagerie of natural astrophysical objects also appear. These include the dusty, gaseous cocoons that hide infant stars and, at the other end of the spectrum, Sun-like stars nearing the end of their lives and exhaling their outer layers. The discarded shell then cools to form a veil of dust that absorbs much of the star's light and re-radiates it at infrared temperatures. Carrigan found about a dozen interesting candidates with infrared emission similar to that expected from a Dyson swarm, but none were particularly convincing, and in the end they were all deemed natural.

Not to be discouraged, in 2015 astronomers led by Jason Wright of Pennsylvania State University revealed the findings from an even more ambitious project called Glimpsing Heat from Alien Technologies, or G-HAT (or Ĝ if you prefer) for short. They weren't looking for just one Dyson swarm. They were searching for a swarm of Dyson swarms: an entire galaxy filled with them. This would be a sign of a Kardashev Type III civilization.

The Soviet astrophysicist Nikolai Kardashev came up with this concept in 1964. An ecumenopolis would be a version of a Type I civilization, one that is able to utilize every scrap of energy available to it on that planet. A Type II civilization harnesses the entire energy output of its star, via a Dyson swarm, and a Type III civilization spreads out among the stars, building Dyson swarms around every one of them in a galaxy.

So G-HAT was searching for entire galaxies that had the specific infrared signature of a Type III civilization. None were found among the 100,000 galaxies (for context, out of several trillion in the observable universe) observed by NASA's WISE (Wide-field Infrared Survey Explorer) space telescope. Can we assume from this that Type III civilizations do not exist? Is this telling us anything profound about the development of technologically intelligent life? Perhaps; all we can say with real confidence is that Type III civilizations are exceedingly rare. Admittedly, G-HAT was unable to rule out galaxies that had only been partially covered in Dyson swarms, so there's that. Rather than hinting at the rarity of life, perhaps the findings are telling us that the Kardashev scale itself is the problem. Remember, it's from the same era as the formal conception of the idea of the ecumenopolis, an era in which growth and consumption were seen as positive drivers of society, rather than being ecologically unsound. Perhaps all these concepts – ecumenopolises, artificial worlds and Kardashev civilizations – are best left to fiction while we try to figure out our own cities.

Not that cities are necessarily bad. Both Arkady Martine and Miguel Kanai have a fondness for them, despite the problems they can have.

'I wanted to write about a city-planet because I love cities,' says Martine: 'I love the ones that feel like they are an organism, and in science fiction it's like we've been given permission to build one that's even bigger, even weirder, even more complicated.'

'I think cities are good,' adds Miguel Kanai. 'Obviously we need very clear land-use policies, but more than controlling urban growth, what we really need is nature protection. I think that's what matters more than trying to restrain cities.'

We may not live in a true ecumenopolis, although the interconnectivity between cities today is able to mimic some aspects of an ecumenopolis, but urbanization, wide-scale industrialization and agriculture are forever changing our world. Their impact upon our planet is ushering in a new geological age, the Anthropocene, in which human activity has the greatest influence on the environment. The breaching of the planetary boundaries is evidence for this. We head into an uncertain future, environmentally, technologically and sociologically. Science fiction promises both splendour and strife, and we would be wise to listen, to pay heed to its warnings and to pursue its treasures if they are to be to the benefit of all life on Earth.

# EPILOGUE

Towards the end of the writing of this book, as summer ticked over into autumn in 2023, something interesting happened.

Light from a faraway star named K2-18 had travelled 124 light years to fall upon the gold-plated beryllium segments of the main mirror of the JWST, 1.5 million kilometres from Earth. That light, by now more than a century old, bounced around the JWST's optical system before coming to rest within the telescope's Near-Infrared Spectrometer. Here the light was split into its component wavelengths to be arranged into a spectrum. At wavelengths corresponding to carbon dioxide and methane, the spectrum dropped out, a dark line where there was no light. These were the signs of atmospheric molecules on an exoplanet absorbing K2-18's light as the planet passed in front of and transited the star.

The resulting transmission spectrum was the most detailed spectrum of an exoplanet smaller than Neptune ever captured. It was the first time that organic molecules – one containing atoms of both carbon and hydrogen, which methane does – had been unambiguously detected in an exoplanet's atmosphere. This was an achievement in itself, given the importance of organic molecules to life.

The observations pointed towards K2-18b possibly being a hycean world. We briefly touched on such planets in Chapter Four. The name hycean (pronounced hy-shun) is a portmanteau of hydrogen and ocean; conceived by the University of Cambridge's

Nikku Madhusudhan, hycean worlds were purely hypothetical, warm mini-Neptunes with deep atmospheres of hydrogen wrapped around a potentially habitable global ocean on top of a core of rock and possibly exotic, dense ices. The spectrum of K2-18b lacked ammonia, which is the strongest evidence yet for the existence of hycean worlds. On Uranus and Neptune, vertical circulation brings ammonia to the cloud tops from deeper within those planets; on a hycean world, the ocean should form a barrier, preventing this. The lack of ammonia in K2-18b's spectrum is therefore quite telling, though admittedly not everybody agrees with the conclusions, and there is still debate among members of the community about just what kind of planet K2-18b is.

Nevertheless, the first possible evidence for hycean worlds is fascinating, for sure, but something else was present. Lurking at wavelengths of 3.5–4.0 microns and 4.5–5.0 microns, was a faint dip that, if real, would correspond to absorption in the planet's atmosphere by a molecule called dimethyl sulphide. The results are tentative, the JWST having only observed two transits of the orbiting planet. The dip may be nothing – spurious data, a bit of noise from activity on the parent star. Or the absorption line could be real, and that could potentially be a really big deal. Because on Earth, the only thing that produces dimethyl sulphide is life, mostly phytoplankton in Earth's oceans.

I told you something interesting happened. Could this be the first detection of a real biosignature, of alien life itself?

I spoke to Madhusudhan, who besides coming up with the idea for hycean worlds in 2021 also led the research team that made this discovery with the JWST. He remains grounded, and cautious – understandable given that it's even money whether further data will cement the evidence for dimethyl sulphide or reveal it to be a spectral ghost. But he explained the possible significance: 'Other scientists have predicted that if you have a hydrogen-rich atmosphere you could have dimethyl sulphide as a biosignature, so if we go by the current literature this should be a pretty good indicator of life.'

However, we're not talking about Earth. We're talking about an alien world unlike any we have seen in our solar system. On a world almost nine times the mass of Earth, and 2.6 times the diameter, with a thick hydrogen atmosphere encapsulating a global ocean, we really don't know for sure what chemical processes can take place. Madhusudhan warns that laboratory experiments hint that ultraviolet light incident on a mix of methane and hydrogen sulphide can produce dimethyl sulphide under perfect conditions.

'Could that happen on a hycean world?' he asks: 'We don't know.'

This is the beauty of being on the cutting edge of a wave of discovery, the ability to find something and say, 'we don't know – but we're going to find out.' Future observations with the JWST, which may well have occurred by the time you read this, should at least determine whether the dimethyl sulphide signature is real. If it is, then who knows where it will take us. And it's not just K2-18b. It's also the four other candidate hycean worlds that the JWST is following up on. And the TRAPPIST-1 planets. And Teegarden's Star b, and Proxima b, and the hunt for habitable exomoons, and myriad more planets rich in possibility. We're only at the beginning of this adventure; SF has been exploring exoplanets for a long time, and only now are astronomers starting to do so for real. We're living in an era where exoplanets are transforming; they are no longer just dips in starlight or a Doppler-shifted wobble, but worlds that feel real in and of themselves, characterized by our telescopes.

Human bias has seen astronomers initially focusing their searches on looking for 'Earth 2.0', a replica of our planet orbiting some other star. But Captain Kirk spoke of 'strange new worlds', habitable planets quite unlike Earth, and K2-18b is certainly both strange and new. Could there really be microbial life in a global ocean on a distant world of a type that nobody had even conceived of until the start of the 2020s? Either way, it seems certain that K2-18b won't be the last strange new planet to snag our attention.

Exoplanet discoveries drive two things home to us. The first is how rare our planet Earth still is. We haven't found another one like it. Temperate, rocky worlds with land and sea and breathable atmospheres may be in short supply. It reminds us that we can't just pick up and leave – we don't have another planet to go to – and we must look after this one with far more care and responsibility than we have done thus far.

The second is that the imagination imbued within science fiction can only carry us so far. Science fiction is wonderfully inspiring, thought-provoking and insightful; through the writing of this book, I've discovered so many new fictional planets in works I hadn't read until now, and found so many exciting authors in the process, and I hope, through this book, you are also able to discover your own strange new worlds and wonderful authors, TV shows and cinema. But SF is not the real deal. No matter how bizarre planets are in science fiction, astronomical history has shown that the universe can throw at us planets even more bizarre than anything we could have dreamed of. Sometimes, truth really is stranger than fiction.

And we're here and waiting to take in all of it.

# BIBLIOGRAPHY

## BOOKS, SHORT STORIES AND ARTICLES

Adams, Douglas, *The Hitchhiker's Guide to the Galaxy* (London, 1979)
Aldiss, Brian, *Helliconia Spring* (London, 1982)
—, *Helliconia Summer* (London, 1983)
—, *Helliconia Winter* (London, 1985)
Anders, Charlie Jane, *The City in the Middle of the Night* (New York, 2019)
Asimov, Isaac, 'Nightfall' (New York, 1941)
—, *Lucky Starr and the Big Sun of Mercury* (New York, 1956)
—, *Foundation* (Herts, 1976)
Balmer, Edwin, and Philip Wylie, *When Worlds Collide* (New York, 1933)
—, *After Worlds Collide* (New York, 1934)
Baxter, Stephen, *Titan* (London, 1998)
—, *The Science of Avatar* (London, 2012)
—, *Proxima* (London, 2013)
—, *The Massacre of Mankind* (London, 2017)
—, *Galaxias* (London, 2021)
—, and Terry Pratchett, *The Long Earth* (London, 2012)
Bova, Ben, *Titan* (London, 2006)
Bradley, Marion Zimmer, *The Planet Savers* (New York, 1958)
Brin, David, *Startide Rising* (New York, 1983)
Burroughs, Edgar Rice, *Under the Moons of Mars* (Chicago, IL, 1912)
Cavendish, Margaret, *The Blazing World* (1666)
Chambers, Becky, *To Be Taught, if Fortunate* (London, 2020)
Clarke, Arthur C., *2010: Odyssey Two* (New York, 1982)
—, *Songs of Distant Earth* (London, 1986)
Clement, Hal, *Mission of Gravity* (New York, 1954)
—, *Close to Critical* (New York, 1964)
Cohen, Jack, and Ian Stewart, *Evolving the Alien: The Science of Extraterrestrial Life* (London, 2002)

Corey, James S. A., *Leviathan Wakes* (London, 2011)
—, *Tiamat's Wrath* (London, 2019)
Forward, Robert, *Flight of the Dragonfly* (New York, 1984)
Foster, Alan Dean, and George Lucas, *Star Wars – From the Adventures of Luke Skywalker* (New York, 1976)
Gibson, William, *Neuromancer* (New York, 1984)
—, *Count Zero* (London, 1986)
—, *Mona Lisa Overdrive* (London, 1988)
Hamilton, Peter F., *The Reality Dysfunction* (London, 1996)
—, *The Confederation Handbook* (London, 2000)
—, *Salvation* (London, 2018)
Heinlein, Robert A., *Starship Troopers* (New York, 1959)
Herbert, Frank, *Dune* (Philadelphia, PA, 1965)
Le Guin, Ursula K., *The Left Hand of Darkness* (New York, 1969)
Lem, Stanisław, *Solaris* (London, 1971)
Liu, Cixin, *The Three-Body Problem* (New York, 2014)
McAuley, Paul, *The Quiet War* (London, 2008)
—, *Something Coming Through* (London, 2015)
—, *Into Everywhere* (London, 2016)
Martine, Arkady, *A Memory Called Empire* (New York, 2020)
Niven, Larry, 'The Coldest Place', *If Magazine* (December 1964)
—, *A Gift from Earth* (New York, 1968)
—, *Ringworld* (New York, 1970)
—, *The Smoke Ring* (New York, 1987)
Reynolds, Alastair, *Revelation Space* (London, 2000)
—, *Chasm City* (London, 2001)
—, *Poseidon's Wake* (London, 2015)
Sinclair, Alison, *Blueheart* (London, 1996)
Smith, E. E. 'Doc', *The Skylark of Space* (Providence, RI, 1946)
Suhner, Laurence, *Vestiges* (Nantes, 2012)
—, 'The Terminator', *Nature*, DXLII/7642 (2017), p. 512
Turtledove, Harry, *A World of Difference* (New York, 1990)
Verne, Jules, *Journey to the Centre of the Earth* (Paris, 1864)
—, *From the Earth to the Moon* (Paris, 1865)
—, *Around the Moon* (Paris, 1869)
Voltaire, *Micromégas* (1752)
Weir, Andy, *The Martian* (New York, 2014)
Wells, H. G., *The War of the Worlds* (London, 1898)
Zahn, Timothy, *Star Wars: Heir to the Empire* (New York, 1991)

## FILMS

*65*, dir. Scott Beck and Bryan Woods (Sony Pictures, 2023)
*Akira*, dir. Katsuhiro Otomo (Toho, 1988)
*Alien*, dir. Ridley Scott (20th Century Fox, 1979)
*Aliens*, dir. James Cameron (20th Century Fox, 1986)
*Avatar*, dir. James Cameron (20th Century Fox, 2009)
*Avatar: The Way of Water*, dir. James Cameron (20th Century Fox, 2022)
*Blade Runner*, dir. Ridley Scott (Warner Bros, 1982)
*Caravan of Courage: An Ewok Adventure*, dir. John Korty (ABC, 1984)
*Dune*, dir. David Lynch (Universal Pictures, 1984)
*Dune: Part One*, dir. Denis Villeneuve (Warner Bros, 2021)
*Dune: Part Two*, dir. Denis Villeneuve (Warner Bros, 2024)
*The Empire Strikes Back*, dir. Irvin Kershner (20th Century Fox, 1980)
*Europa Report*, dir. Sebastían Cordero (Magnolia Pictures, 2013)
*Ewoks: The Battle for Endor*, dirs Jim and Ken Wheat (ABC, 1985)
*Forbidden Planet*, dir. Fred M. Wilcox (Metro-Goldwyn-Mayer, 1956)
*Interstellar*, dir. Christopher Nolan (Paramount/Warner Bros, 2014)
*Metropolis*, dir. Fritz Lang (Parufamet, 1927)
*Outland*, dir. Peter Hyams (Warner Bros, 1981)
*Pitch Black*, dir. David Twohy (Universal Pictures, 2000)
*Predator*, dir. John McTiernan (20th Century Fox, 1987)
*Predators*, dir. Nimród Antal (20th Century Fox, 2010)
*Return of the Jedi*, dir. Richard Marquand (20th Century Fox, 1983)
*Star Trek II: The Wrath of Khan*, dir. Nicholas Meyer (Paramount Pictures, 1982)
*Star Wars*, dir. George Lucas (20th Century Fox, 1977)
*Star Wars: Episode I – The Phantom Menace*, dir. George Lucas (20th Century Fox, 1999)
*Star Wars: Episode II – Attack of the Clones*, dir. George Lucas (20th Century Fox, 2002)
*Star Wars: Episode III – Revenge of the Sith*, dir. George Lucas (20th Century Fox, 2005)
*Star Wars: The Rise of Skywalker*, dir. J. J. Abrams (20th Century Fox, 2019)
*Total Recall*, dir. Paul Verhoeven (TriStar Pictures, 1990)
*When Worlds Collide*, dir. Rudolph Maté (Paramount Pictures, 1951)

## TELEVISION SHOWS

*The Ark* (Syfy, 2023–)
*Babylon 5* (Warner Bros Television, 1993–8)
*Doctor Who* (BBC, 1963–)
*The Expanse* (SyFy/Amazon Prime, 2015–22)
*Foundation* (AppleTV+, 2021–)
*Game of Thrones* (HBO, 2011–19)
*The Mandalorian* (Disney+, 2019–)
*Space: 1999* (ITV, 1975–7)
*Star Trek* (NBC, 1966–9)
*Star Trek: The Next Generation* (Paramount Domestic Television, 1987–94)
*Star Trek: Deep Space Nine* (Paramount Domestic Television 1993–9)
*Three-Body Problem* (Netflix, 2024–)
*V* (Warner Bros Television 1983–4)

## COMICS

*2000* AD (London, 1977–)
*Flash Gordon* (New York, 1934–92)
*Judge Dredd Megazine* (London, 1990–)
*Superman* (New York, 1939–)
*The Transformers* (New York and London, 1984–92)

# ACKNOWLEDGEMENTS

I would like to thank the following people who helped me with the science-fiction side of things: Charlie Jane Anders, Stephen Baxter, Paul McAuley, Arkady Martine and Alastair Reynolds. They all kindly made time for me and allowed me to interview them.

From the academic community, I would like to thank Jessie Christiansen, Knicole Colón, Dale DeNardo, Alex Farnsworth, Claire Guimond, René Heller, Alex Howe, Juan Miguel Kanai, Grant Kennedy, Nikku Madhusudhan, Abel Méndez, Claire Newman, Adiv Paradise, Emma Puranen, Sara Seager, Amaury Triaud and Robin Wordsworth.

I would also like to thank those who helped *Amazing Worlds* along the road to publication: Michael Leaman, Amy Salter, Fran Roberts, Helen McCusker, Alex Ciobanu, Miren Lopategui, Phoebe Colley and the rest of the team at Reaktion Books.

# PHOTO ACKNOWLEDGEMENTS

The author and publishers wish to express their thanks to the sources listed below for illustrative material and/or permission to reproduce it:

ALMA (ESO/NAOJ/NRAO)/Balsalobre-Ruza et al.: p. 120; ESA/DLR/FU Berlin/NASA MGS MOLA Science Team: p. 115 (*top*); ESO: pp. 114 (*bottom*), 115 (*bottom*; M. Kornmesser), 117 (P. Kervella (CNRS/U. of Chile/Observatoire de Paris/LESIA), Digitized Sky Survey 2/D. De Martin/M. Zamani), 118 (*top*; L. Calçada); Benoît Gougeon, Université de Montréal: p. 119 (*bottom*); NASA/JPL-Caltech: pp. 116 (R. Hurt (IPAC)), 118 (*bottom*), 119 (*top*); NASA's Ames Research Center: pp. 114 (*top*); NRC-HIA/C. Marois/Keck Observatory: p. 113.

# INDEX

Page numbers in *italics* refer to illustrations